UNLOCKING THE GLOBAL WARMING TOOLBOX

Key Choices for Carbon Restriction and Sequestration

Steven Ferrey

> **Disclaimer:** The recommendations, advice, descriptions, and the methods in this book are presented solely for educational purposes. The author and publisher assume no liability whatsoever for any loss or damage that results from the use of any of the material in this book. Use of the material in this book is solely at the risk of the user.

Copyright © 2010 by
PennWell Corporation
1421 South Sheridan Road
Tulsa, Oklahoma 74112-6600 USA

800.752.9764
+1.918.831.9421
sales@pennwell.com
www.pennwellbooks.com
www.pennwell.com

Marketing: Jane Green
National Account Executive: Barbara McGee

Director: Mary McGee
Managing Editor: Stephen Hill
Production Manager: Sheila Brock
Production Editor: Tony Quinn
Book Designer: Susan E. Ormston
Cover Designer: Alan McCuller

Library of Congress Cataloging-in-Publication Data

Ferrey, Steven.
 Unlocking the global warming toolbox : key choices for carbon restriction and sequestration / Steven Ferrey.
    p. cm.
  Includes bibliographical references and index.
  ISBN 978-1-59370-213-7
 1. Global warming--Prevention--Government policy. 2. Carbon sequestration--Government policy. 3. Air quality management--International cooperation. 4. Greenhouse gas mitigation--International cooperation. 5. Environmental law. I. Title.
 GE170.F47 2010
 363.738'74--dc22
                                                                    2009050914

All rights reserved. No part of this book may be reproduced, stored in a retrieval system, or transcribed in any form or by any means, electronic or mechanical, including photocopying and recording, without the prior written permission of the publisher.

Printed in the United States of America

1 2 3 4 5   14 13 12 11 10

CONTENTS

Part I Industrialization and Carbon in the 21st Century: Examining the Chemistry, Location, Timing, and Future of Warming 1

    1 Opening the Toolbox and What It Offers. 3

    2 The Science Underlying Global Warming. 13

    3 The Critical Role of Electric Power Architecture and Carbon 27

    4 The Tipping Point: Time as the Enemy 41

Part II Cap-and-Trade Carbon Regulatory Mechanisms in Place across the World 49

    Section 1: Europe and the World

    5 The Kyoto Protocol: The World Carbon Model 51

    6 The European Union Core of Carbon Control: Compared and Contrasted with Recent U.S. Experience..... 61

    Section 2: The United States

    7 The Regional Greenhouse Gas Initiative: The Original U.S. Regulation. 79

    8 Golden State Carbon: California GHG Regulation 91

    9 Regional and Voluntary U.S. Carbon Programs. 103

Part III The Legal and Policy Issues Confronting Carbon Control Worldwide: Manipulating the Toolbox of Regulatory Options.................................... 111

    10 The Kyoto Critique: The Urgency of International Redesign 113

    11 The Fulcrum Leverage on Global Warming: Role of the Courts 133

    12 The New Carbon-attuned Smart Grid: Beyond Simple Poles and Wires 147

    13 Carbon Leakage and the Commerce Clause 167

    14 Carbon Allowance Auction: Regulatory and Legal Issues... 187

    15 Legal Additionality Requirements for Carbon Offsets..... 203

Part IV Carbon Regulation Interfacing with Renewable Power: Renewable Tools from the Toolbox 215

    16 Offsetting Carbon: Creating Credits from Renewable Power and Conservation 217

    17 The Feed-in Tariff for Renewable Energy: Where It Works and Where It Encounters Legal Impediments 229

    18 Renewable Portfolio Standards for Renewable Power 247

    19 The Successful Architecture to Transform Renewable Power 263

    20 Into the Woods 277

    21 The Final Analysis: The Conclusion on Carbon........... 289

Appendix: Abbreviations 299

Index 303

PART I

INDUSTRIALIZATION AND CARBON IN THE 21ST CENTURY:
EXAMINING THE CHEMISTRY, LOCATION, TIMING, AND FUTURE OF WARMING

1 OPENING THE TOOLBOX AND WHAT IT OFFERS

Energy is the biggest business in the world, with sales each year of about $2 trillion.[1] As the reader opens this book, the United States will be deciding how, and whether through state governments, the federal government, or both, to regulate global warming and climate change for our, and our children's children's futures. Because CO_2, once emitted, remains warming the atmosphere for a century after its release, none of us reading this book will see those molecules of carbon to their final warming end. Therefore, it is critically important to get the model of implementation right from the beginning. This is a regulatory toolkit holding some of the policy and legal implements necessary to sculpt the carbon future.

This book is different from others in two important dimensions. First, most literature on climate change tracks the latest pending legislation. This book does not follow that script for several deliberate and conscious reasons:

- The Waxman-Markey bill has become the 2009 vehicle for legislative debate, and in the period of a month, in the House of Representatives it changed from 600 pages in length to more than 1,400 pages, as accommodations were made to various interests. It is very different from the Lieberman-Warner bill that was the vehicle for carbon control in the Congress in 2008. What a difference a year makes! The current legislation must pass the Senate and be reconciled in a conference committee before it is operative. Then, because of its billions of dollars of impact on the economy and amendments that are longer than the original bill itself, it will continue to be amended and fine-tuned even after enactment. It has a long way to go, and is years away from implementation.

- Regulations written by various federal agencies, including the Environmental Protection Agency, the Federal Energy Regulatory Commission, the Department of Agriculture, the Commodity Futures Trading Commission, and others, will dictate the final shape and impact of carbon regulation at the federal level. These are not yet promulgated or in place.
- The Waxman-Markey legislation to implement U.S. carbon regulation will not be effective until 2012 under its own terms, long after this book was published. There will be regular articles in the press about how this program of carbon regulation is maturing prior to its implementation, and that moving target is better captured in articles in the years ahead rather than in this toolbox.

Instead, this book showcases the transcendent and timeless legal and regulatory issues that will continue for years to be worked out both in the United States and in international programs. Second, this book does not rely on any current snapshot of fast-evolving carbon regulation. Instead, this book tunes into the lasting issues to craft successful carbon control. Nonetheless, in this first chapter we examine the basic contours of the upcoming federal U.S. carbon regulation, as this provides context for the future use of the toolkit tools.

This book is very much a toolbox for the future, and not dependent on the carbon plan du jour. This toolbox highlights key climate control issues now confronting the Obama administration's carbon control program, as well as those of the European Union and the Kyoto Protocol. Half of the chapters that follow examine these transcendent issues which will continue to be confronted over the next decade as carbon regulation matures:

- What greenhouse gases to regulate (chapter 2)
- How urgently to implement carbon restrictions (chapter 4)
- Design of a smart grid (chapter 12)
- Leakage of carbon into the control area (chapter 13)
- Preemption of state carbon programs and auction of allowances (chapter 14)

- What to allow as additional offset credits for carbon control (chapter 15)
- The use of renewable power options to control carbon emissions (chapter 16)
- Use of feed-in tariffs as a regulatory technique (chapter 17)
- How to implement renewable energy portfolio requirements (chapter 18)
- The role of forests and biologic sequestration (chapter 20)

In fact, as the reader opens this toolbox, the major European countries will be assessing why their global warming control has resulted in more, rather than less, carbon emissions:

- The Kyoto Protocol has not been successful in achieving its objectives:
    - Reducing developed countries' emissions of CO_2
    - Promoting sufficient renewable power infrastructure in developing countries
    - Reducing world global warming gases sufficiently
- The 23 states in the United States beginning carbon regulation became ensnarled in legal problems:
    - Preventing renewable energy projects from creating tradable offsets
    - Running potentially afoul of constitutional limits on permissible actions
    - Discriminating against power moving at the speed of light in interstate commerce

Simultaneously, the 190 world countries that ratified the international Kyoto Protocol a decade ago are locked in disputes over how, and even whether, that protocol continues at all past 2012. And if so, which countries must shoulder what burden to implement a radical reduction in the use of fire to manipulate the universe. As a world society, for thousands of years, we have utilized fire to supply much of the energy and power on which the world is now constructed. Now with diminishing supplies of traditional fuels and the necessity to quickly implement climate controls on the global temperature, new challenges confront a seven-billion-person world.

Collectively, every developed nation of the world must refashion its use of power and energy. There is a collective obligation, as every global warming gas molecule, no matter who released it, torques the thermostat of the planet. Unlike other air, water, and land emissions, the impact of greenhouse gases (GHGs) is not distinctly local or regional. It is not coincidental that global warming gases are global. Impact is universal. Therefore, the solution ultimately must also be global.

The dimension of time is not precisely known. Leading scientists warn that there is less than a decade to dramatically reduce GHGs—not to agree on a plan, but actually have a dramatic carbon reduction plan operational and effective. Whether these predictions will prove accurate or not, it is clear that humankind has already significantly affected climate change, and that the pace of the warming effect has accelerated each year. The world is past the time of experimentation, regarding significant changes necessary in the production and use of power in society.

While the chapters that follow take a holistic look at different aspects of, and tools for, global warming policy, it is particularly an issue of power. Electricity has become the signature technology of the 21st century. Access to power separates the "haves" from "have-nots" on the geopolitical map of the globe. A night space satellite photograph of the earth distinguishes this reality, as shown in figure 1-1.

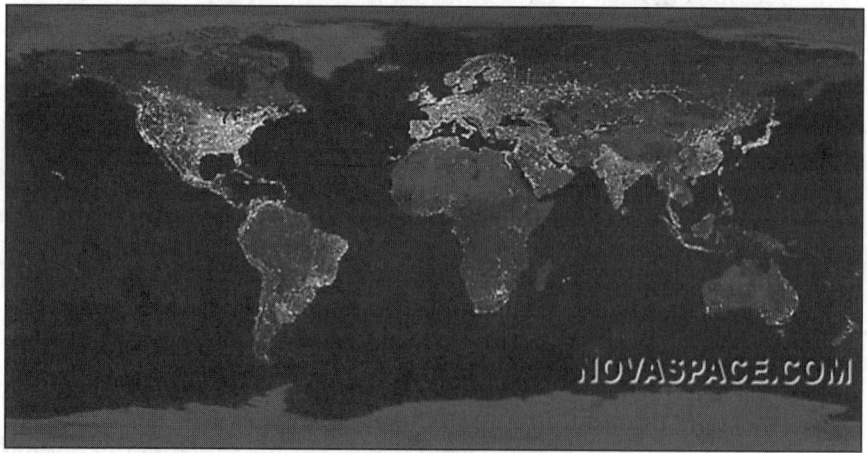

Fig. 1–1. Access to electric power is not distributed evenly across the globe.

Abundant electric power is the cornerstone of the near-term plan of every developing nation in the world. My past 15 years of work as legal advisor to the World Bank and the United Nations Development Programme in developing nations across the globe has made me very aware of the particular underappreciated choices for increasing power supply in developing nations.[2] The risk of energy policy failure carries profound scientific implications. A full regulatory and legal toolbox is needed at our disposal.

A Quick Overview of the Pending U.S. National Carbon System

Between now and 2012, the Waxman-Markey legislation can become the vehicle for addressing carbon at the federal level during the Obama administration. It is relevant to understand its basic structure as an important element of the future, although each of the tools for the evolving system, as well as for the Kyoto Protocol and the 23 state programs in the United States regulating carbon, appear in the various chapters that follow. Among the primary provisions in its 1,400 pages of text, the Waxman-Markey legislation regulates carbon emissions through a cap-and-trade system, creates a national requirement for the increased use of renewable energy, and promotes a smart grid for transmission of electric power.

The Waxman-Markey carbon program would reduce annual emissions by 3% by 2012 against a 2005 baseline, 17% by 2020, and 83% by 2050. It would do so by regulating emitters of 25,000 tons of carbon dioxide equivalent (CO_2e) global warming chemicals in selected industries, while exempting forestry, agriculture, and other industries from the cap. It would do so by the auction or free allocation of tradable emission allowances, with specific expenditure goals for the billions of dollars expected to be garnered from these auctions annually. The bill was changed from the initial auction of most allowances, to the free allocation of most allowances initially, with the number of free allowances being phase out in favor of greater auction of allowances between 2027 and 2030. Initially, 85% of allowances will be allocated for free to distribution utilities, merchant power generation facilities, steel, iron, paper, cement, refinery, and other competition-sensitive businesses. By 2030, 70% of allowances would be auctioned. The tools to determine the free allocation of hundreds of billions of dollars of free allowances are controversial and dealt with in the chapters that follow.

The formula for allocation of the number of free allowances to industrial emitters of greenhouse gases has become extremely controversial in distributing these hundreds of billions of dollars of free allowances. Also controversial is whether the formulae for distribution within an industry will be based on industry averages, specific historic emissions of a particular company, volume of sales, or a hybrid combination of those tools. These allowances could be traded, with some specific companies forecast to receive as many as $1 billion annually of surplus free allowances above and beyond what they required, which they could sell. So-called "early reduction" credits can be earned for carbon reductions made between 2001 and 2009. This creates value in previously unmonetized voluntary CO_2 reductions, as discussed in the chapters that follow. For some smaller emitters, the bill would mandate implementation of carbon reduction technology requirements for the emitting sources, but not require obtaining allowances for compliance regarding emissions. State carbon control programs in 23 U.S. states, described in the following chapters, could be preempted for a temporary five-year period. Each of the various tools for such mechanisms of existing world carbon control is examined in the following chapters.

Compliance will be able to be met with *offsets* up to almost one-third of the number of required allowances. Offsets are additional compliance credits created after 2008 by reducing carbon at locations other than the regulated emitters. Thus, entrepreneurs can reduce carbon at a lower-cost location, create and register an offset credit, and then trade that monetized credit to those trying to find the most cost-effective way to cover their carbon emissions in the regulated system. As examined in the chapters that follow, there is significant controversy about the efficacy of offsets, their verification, and where and how these tools can be created and traded.

Regarding renewable energy, the Waxman-Markey legislation would require 6% of retail electricity to come from renewable power sources by 2012, and 20% by 2020. What qualifies as renewable is a controversial subject, as will be explored in chapters that follow. This U.S. system will include an additional requirement to deploy energy efficiency measures to meet 1% of requirements by 2012 and 15% by 2020. These eligible technologies will have to be defined in regulations. Energy efficiency investments constitute an important component of shifting the new electric grid to a more sustainable metric. The almost 30 existing state programs requiring renewable energy in the retail electric portfolios of power sellers will not be preempted by upcoming federal programs. Therefore, there will be differences in renewable and efficiency requirements and what qualifies state by state. These state programs are discussed in the chapters that follow.

Having provided this brief overview of where the U.S. system of carbon regulation is headed for implementation beginning in 2012, the material in this book is not dependent on the eventual shape of still evolving federal programs. Rather, it examines the basic tools and issues accompanying carbon controls both in the U.S. states and worldwide. Europe has five years of experience and history of carbon regulation, and 10 U.S. states even have a year of experience with carbon regulation. It is in this history and experience that the transcendent issues emerge. And from that experience, the tools are fashioned in this toolbox.

The Toolbox Tools

This toolbox focuses on sectors and solutions. Technologies exist in proven application to make an energy transition. It is not a technological conundrum that confronts us—it is the challenge to forge appropriate policy in real time. A transition will require the concerted efforts of policymakers, regulators, industry, nongovernmental organizations (NGOs), and consumers. It truly is global.

This book is a toolbox for those who would regulate carbon and those who would be subject to that regulation. It critiques each of the carbon regulatory schemes in the world at the beginning of the second decade of the 21st century, assessing what is working right and what is malfunctioning with each element. The fissures in the world Kyoto Protocol are analyzed, as are issues with U.S. carbon regulation. Both the emission sources—with a special emphasis on the utility sector—and carbon absorption and conversion by natural forces and forests, are in the toolkit.

The third section of this toolbox focuses on pivotal legal issues of climate change:

- What has gone wrong with the Kyoto Protocol, and how it must be reformed to have any possibility for long-term success
- The new green grid and its implications
- The important issue of border leakage and the U.S. Constitution's Commerce Clause
- Auction of carbon allowances, the U.S. Constitution's Supremacy Clause, and legal preemption

The toolbox is organized into four sections:

1. An analysis of the science and timing of climate change, as well as the critical role of the electric power industry in any viable solution
2. A detailed examination of how carbon regulation is addressed differently in the Kyoto Protocol involving 190 ratifying world nations, the European Union regulation of carbon in 27 developed nations, and the regulation of carbon in 23 U.S. states under four different regulatory systems
3. What has gone legally wrong torquing carbon policy in the Kyoto Protocol, the EU carbon scheme, and U.S. carbon regulation, with particular emphasis on the following:
   - Allowance allocation and carbon auction
   - Leakage of carbon across borders
   - Regulating carbon upstream or downstream in the power industry
   - The role of the new smart grid
   - The new requirement of *additionality* in carbon offset certification
4. The interface of renewable power technologies and carbon control:
   - The role of renewable portfolio standards
   - The alternative of renewable feed-in tariffs
   - The role of forest preservation and natural resources in carbon control
   - The successful world model for development of renewable energy

Cap-and-trade is the regulatory mechanism through which world economies have decided to regulate carbon emissions. Cap-and-trade is the establishment of emissions limits on certain sources, allocation or sale of the legal rights to emit, and the ability of entities to trade for more or less quantity of such allowances. It is the policy alternative to a tax on carbon use or emissions. U.S. Energy Secretary Steven Chu announced that he and President Barack Obama support a simple cap-and-trade system for the United States, which would "integrate" with the systems in the European Union.[3]

But we live in the best of times and in the worst of carbon times. Neither cap-and-trade nor a broad new tax is readily accepted by all stakeholders. The world economic crisis is causing developing areas to second-guess the rate and means of regional GHG reduction. Regional reliance on coal generation has become a sticking point in reducing carbon, both between states and even among differently positioned electric utilities within states.

Auction of allowances is controversial. At issue is whether allowances to emit carbon should be auctioned to highest bidders or allocated to carbon sources without charge,: If allowances to the power sector are allocated, should the allocation metric be determined by historic emissions (which reflects the historic carbon intensity of particular fuels at each source), gross power sales (which ignores the composition of the carbon intensity of generation at each source), or gross revenues from power sales (which could reflect differences among traditionally regulated and restructured utilities, as well as investor-owned or public utilities)?

Stakeholders have threatened or already initiated suits against early adopting East Coast Regional Greenhouse Gas Initiative (RGGI) states, California, and the European Union regarding these brewing disputes. Disputes have occurred between sovereign states and within states, pitting electric utilities and power suppliers against each other. In the 10 U.S. East Coast RGGI states, a suit by Indeck-Corinth against New York's RGGI program was settled by New York, giving the plaintiffs everything they sought in order to prevent the court from reaching the issue of whether RGGI was legal in New York. Ultimately, electricity is more than just the technology of power. Electricity is the signature energy form of the 21st century global economy.

Even where there appears unanimity on climate change, there is less consensus than at first meets the eye. The European Union Emission Trading Scheme (EU-ETS), operating since 2005, includes 85% of all world countries now regulating their carbon emissions. Chapter 6 reveals the pattern of recently escalating EU disputes, which is a prologue for similar schisms that will appear and mature as the United States attempts to craft effective carbon regulation.

Moreover, in times of economic recession, there is worldwide pressure for the carbon regulatory system to morph into a revenue-raising scheme through auction of emission allowances, in contrast to all prior cap-and-trade emission control systems in the world. Contraposed to this element is the possible ecological *tipping point* forecast within approximately five years. The failure of aggressive global warming policy could tip the world into potentially catastrophic and irreversible consequences from warming.

The policy and regulatory choices made now will sculpt the world response on what has been called the environmental challenge of this century. There is cause for both great optimism, as well as concern. The technologies exist today to change the power generation base. However, since the concept of global warming control was agreed among world powers in 1992, little has been achieved amid a four-fold increase in the rate of global warming gases pouring into the atmosphere.

It is time to examine and choose the correct tools, and get about serious business on climate change. The toolbox opens in the next chapters. Roll up your sleeves and open the part of the toolbox of most interest.

Notes

1 United Nations Conference on Trade and Development. 2001. *Energy Services in International Trade: Development Implications.* TD/B/COM.1/EM.16/2, June 18, sec. 3.

2 Ferrey, Steven and Anil Cabraal. 2006. *Renewable Power in Developing Countries: Winning the War on Global Warming.* Tulsa: PennWell.

3 Carbon Control News. 2008. New DOE Secretary Backs Cap-and-Trade. *carboncontrolnews.com.* January 13.

2) THE SCIENCE UNDERLYING GLOBAL WARMING

It's Getting Hotter

Prior to the Industrial Revolution, average Earth temperature had been naturally maintained at 59°F. Since the Industrial Revolution, carbon emissions resulting from combusting fossil fuels to provide mechanical and electrical energy have poured into the atmosphere, as shown in figure 2–1.[1] Current atmospheric CO_2 levels of about 382 parts per million (ppm) are approximately 33% higher than in preindustrial times. For the past 50

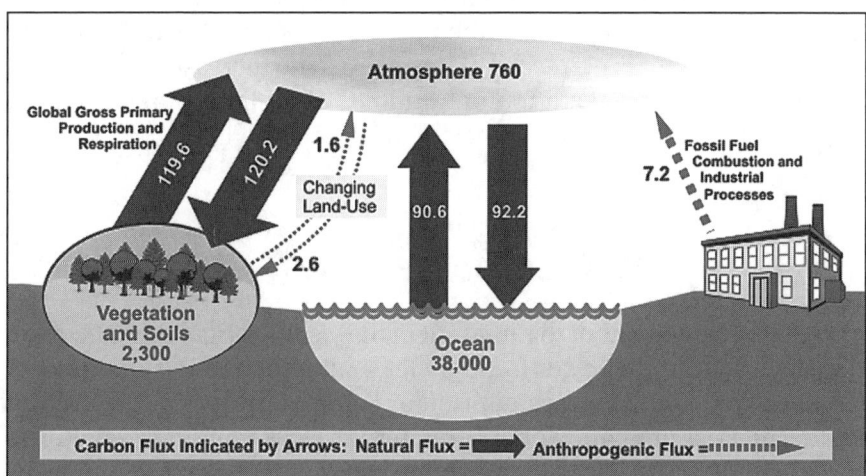

Fig. 2–1. The global carbon cycle displaying flow of carbon into air and water systems. *Source: Intergovernmental Panel on Climate Change, Climate Change 2007: The Physical Science Basis, Figure 7.3 and Table 7.1. (U.K., 2007)*

years, scientists have known about the warming effects of carbon dioxide. Temperature changes move in direct relation to atmospheric greenhouse gas (GHG) concentrations.[2]

Context is important: Carbon-induced global warming has played an essential role in human history on this planet. Early in the billions of years of Earth's formation, the original atmosphere consisted of carbon dioxide, nitrogen, methane, and sulfur. The greenhouse effect of the carbon dioxide concentrations in the atmosphere was essential to the basic warming of the planet. In the early days of the earth, the sun was significantly dimmer, and this warming was essential to raise Earth's temperature to a level capable of sustaining human life.[3] Without this carbon dioxide, the Earth could have frozen permanently.

Limiting modern era global warming to an increase of 4°F, will require stabilizing CO_2 concentrations in Earth's atmosphere at no more than 450 ppm. A top official with the United Nations Intergovernmental Panel on Climate Change (IPCC) has indicated that developed nations will need to slash CO_2 emissions by 80–95% (or almost entirely) by 2050 to hold GHGs concentrations to 450 ppm in the atmosphere.[4] CO_2 concentrations in the atmosphere remain for decades, or even centuries. Within a century, if all nations of the world do not limit greenhouse gas emissions, the average global temperature will climb anywhere from 1.4° to 5.8°C (or 2.5° to 10°F).[5]

Consider the context: For the past 10,000 years, the Earth's temperature has varied by less than 2°F. Global mean surface temperature rose 1.33°F over the last decade, and the rate of warming over the past 50 years has almost doubled. Eleven of the past 12 years have been among the warmest dozen years on record. Such an extreme increase (10°F) would not only lead to the starvation of hundreds of millions of persons, but usher in the mass extinction of half of the species on Earth.[6]

Greenhouse Gases

GHGs are transparent to radiation in the visible part of the spectrum, but absorbent in the lower frequencies, including the infrared part of the spectrum. Thus, they retain outgoing radiation trying to leave the earth, and raise ambient global temperature. Heat-trapping greenhouse gases include: water vapor, carbon dioxide (CO_2), methane (CH_4), nitrous oxides (N_2O), sulfur hexafluoride (SF_6), hydrofluorocarbons (HFCs), and perfluorocarbons (PFCs).[7]

The Kyoto Protocol chose to regulate six families of global warming chemicals, but not all. The regulated heat-trapping greenhouse gases do not include water vapor. The impact of water vapor is to raise the temperature of Earth enough so that it is warm enough to be habitable.[8] Fossil fuel generation produces 64% of the total atmospheric CO_2, and this amount has increased significantly since 1990.[9]

The GHGs in table 2-1 are displayed in descending order of their impacts on the environment, which is a function of their quantity released, heat radiation properties, and residence time in the atmosphere.

Table 2-1. Key facts about greenhouse gases

GHG	Global Warming Potential [CO_2=1]	Residency Time [years]	Amount of U.S. Total GHG Release [%]
Carbon dioxide (CO_2)	1	100	85
Methane (CH_4)	21	12	11
Nitrous oxides (NO_x)	310	120	2
Hydrochlorofluorocarbons (HFCs)	140–11,700	Varies	<1
Chloroflurocarbons (CFCs)	6,500	Varies	<1
Sulfur hexafluoride (SF_6)	23,900	Varies	<1

The molecule-by-molecule global warming impact of many of these secondary and less prevalent GHGs is significantly greater than CO_2. However, because these secondary GHGs are released in smaller quantities and/or have shorter residence times in the atmosphere before they dissipate, CO_2 remains the dominant GHG and therefore receives the greatest international policy focus. This relative impact is illustrated in figure 2-2.

The majority of Kyoto Protocol Clean Development Mechanism (CDM) offset projects (see chapters 7 and 10) have been projects to reduce HFC-23, a carbon-based refrigerant. These offset projects address high global warming potential (GWP) industrial gases such as trifluoromethane (HFC-23) and N_2O as well as CH_4 emitted by landfills and concentrated-animal-feeding operations (CAFOs). Two relatively obscure industries—adipic acid and chlorodifluoromethane (HCFC-22)—dominate CDM projects. Adipic acid is the feedstock for the production of nylon-66 and releases abundant N_2O as a production by-product. HCFC-22 has two major applications. It is one of two major refrigerants that were phased in to replace the CFCs under the Montreal Protocol on Substances that Deplete the Ozone Layer. HCFC-22 is also the primary feedstock in the production of DuPont Teflon. These two relatively small industries represent nearly 55% of the supply of issued certified emission reductions (CERs) in the CDM to date.

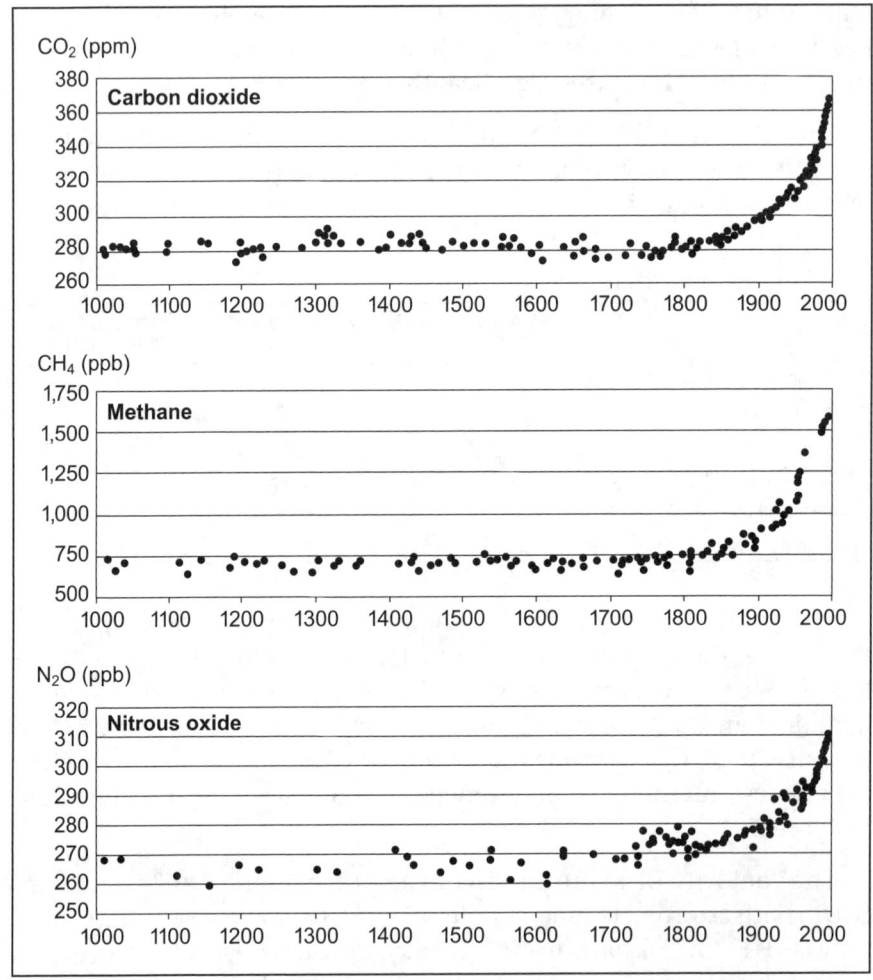

Fig. 2–2. Greenhouse gas emissions 1980–2006 as reported in the Energy Information Administration's Annual Energy Review 2007

Indeed, these industrial gas emissions do not even occur in the developed world, not because of an absence of adipic acid or HCFC-22 manufacture, but because industries abated them voluntarily and destroy them.[10] A developing world producer of HCFC-22 can earn more than twice as much from its CDM offset subsidy as it can gross from the sale of its primary product, tripling revenues and profits. The relative impact of GHGs is illustrated in figure 2-3.

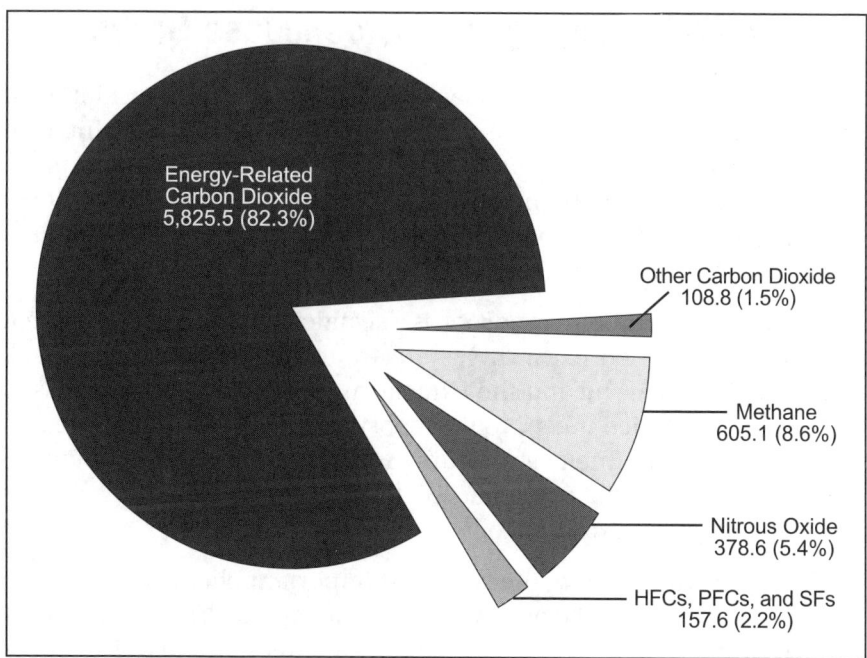

Fig. 2–3. U.S. anthropogenic greenhouse gas emissions in 2006 (million metric tons of carbon dioxide equivalent). *Source: Energy Information Administration, Emissions of Greenhouse Gases in the United States 2006 (Washington, D.C., November 2007)*

HFCs constitute less than 1% of GHGs, but they have received almost half of Kyoto CDM value of investment in mitigation of GHG. This disparity occurred because of the 11,700 times greater CO_2e value of a molecule of HFC; for an investment of €100 million, it generates CDM credit revenues worth €4.6 billion. A 46-fold return on investment is more than impressive even in conventional investments. Ironically, HFC emitters earned almost twice as much from these CDM HFC by-product reduction activities than they did from selling their refrigerant gases produced as the salable commodity in the process. HFC capture and destruction projects do nothing to shift the energy base of the world's economies to sustainable, renewable power technologies. The necessary structural energy sector transformations to low-carbon generating technologies are not occurring at anywhere near the pace of HFC reductions.

Forgotten and Ignored Greenhouse Gases

Black carbon (BC) particulate emissions has been recently identified as the second most important climate changing agent, trapping heat as an aerosol, and changing albedo of snow and ice.[11] "[E]missions of black carbon are the second strongest contribution to current global warming, after carbon dioxide emissions," according to V. Ramanathan and G. Carmichael.[12]

Black carbon, also known as soot, is a significant but underappreciated GHG emission of concern. BC is an important warming chemical, especially in the Arctic and the Himalayas, especially affecting loss of the West Antarctic ice sheets. "Soot deposition increases surface melt on ice masses, and the meltwater spurs multiple radiative and dynamical feedback processes that accelerate ice disintegration," according to James Hansen and Larissa Nazarenko.[13]

If true, the irony is that the two most important chemical emissions affecting warming are not covered by the Kyoto Protocol or other international global warming regulation mechanisms. The most significant global warming gas, water vapor, is not covered under any global warming regulatory schemes. A synthetic gas manufactured by industry to make plasma screens and flat panel televisions, nitrogen trifluoride (NF_3), is the second most potent warming gas after SF_6, but is not regulated by any international carbon restriction laws. Black carbon is not covered in any way under the Kyoto Protocol, nor did it appear on the 2007 Kyoto Protocol Bali agenda for future reforms.

However, this omission of black carbon is of particular significance. Regulation of black carbon "particularly from fossil fuel sources, is very likely to be the fastest method of slowing global warming" in the immediate future, according to Mark Jacobson at Stanford University, who notes that major cuts in soot emissions could slow the effects of climate change for a decade or two.[14] Jacobson also notes it would give policymakers more time to address CO_2 emissions.[15] This is an extension of global warming tools that policymakers will require given lack of progress on global warming progress over the decade since the adoption of the Kyoto Protocol. Jacobson states that black carbon from fossil fuel and biofuel soot "may contribute to about 16% of gross global warming, but its control in isolation could reduce 40% of net global warming minus cooling from all negative radiative forcings (e.g., sulfates)."[16] What makes this complex to factor into the GHG equation is that reducing aerosols that are not black carbon could actually increase global temperatures by up to 2.4°C.[17]

Of the black carbon emission total, about 20% is emitted from burning biofuels, 40% from fossil fuels, and 40% from open biomass burning. Ramanathan estimates that "providing alternative energy-efficient and smoke-free cookers and introducing transferring technology for reducing soot emissions from coal combustion in small industries could have major impacts on the radiative forcing due to soot."[18] Simultaneously, reducing black carbon emissions would save up to three million lives a year that otherwise would be lost to air pollution.[19] There is an international and locational component to such an implementation tool.

Today, unlike the policy focus on CO_2, the overwhelming majority of black carbon emissions is from developing countries[20] and are expected to increase.[21] The largest sources of ambient black carbon emissions are in Asia, Latin America, and Africa.[22] China and India, the two countries with the precipitous increases in deployment of coal-fired power resources, alone account for 25–35% of total worldwide global black carbon emissions, with emissions from China doubling from 2000 to 2006.[23] The black carbon concentration hotspots include "the Indo-Gangetic plains in South Asia; eastern China; most of Southeast Asia, including Indonesia; regions of Africa between sub-Sahara and South Africa; Mexico and Central America; and most of Brazil and Peru in South America," housing about three billion people, or half the world's population.[24]

Developed nations have reduced their black carbon emissions from fossil fuel sources by a factor of five or more since 1950.[25] Attention on the power generation sector in Asia would be one of the most cost-effective ways to reduce emissions of black carbon. Unfortunately, tools addressing black carbon emissions are not contained in the Kyoto Protocol. Developing countries have not made this transition, and will not do so quickly without mechanisms in the international legal protocols.

The Canary in the Carbon Coal Mine

The early evidence is in. The Arctic is the world's early-warning system, warming at a rate twice that of the rest of the world. In the winter of 2004–2005, a chunk of ice equivalent in area to Turkey, simply cracked and fell into the sea.[26] At recent average rates, approximately 40% of the perennial ice cap will have disappeared by the year 2050, long before earlier studies had predicted. At the 2007 rate, it will disappear in less than a decade. In 2007, shattered records for Arctic melt were as follows:

- 552 billion tons of ice melted this summer from the Greenland ice sheet, according to preliminary satellite data. That's 15% more than the annual average summer melt, beating the 2005 record.

- A record amount of surface ice was lost over Greenland in 2007, 12% more than the previous worst year, 2005, according to data from the University of Colorado. That's nearly quadruple the amount that melted just 15 years ago. It's an amount of water that could cover Washington, D.C., a half-mile deep, researchers calculated.

- The surface area of summer sea ice floating in the Arctic Ocean this summer was nearly 23% below the previous record. The dwindling sea ice has already affected wildlife, with 6,000 walruses coming ashore in northwest Alaska in October 2007 for the first time in recorded history. In another first, the Northwest Passage was open to navigation.

- NASA data show the remaining Arctic sea ice to be unusually thin, another record. That makes it more likely to melt in future summers. Combining the shrinking area covered by sea ice with the new thinness of the remaining ice, scientists calculate that the overall volume of ice is half the 2004 total. White sea ice reflects about 80% of the sun's heat off Earth. When there is no sea ice, about 90% of the heat goes into the ocean which then warms everything else up. Warmer oceans then lead to more melting.

Figure 2-4 shows the Arctic ice sheet from 1979 to 2006. The obvious loss of ice is apparent.

The following was noted by Al Gore in accepting the Nobel Peace Prize:

> Last September 21, as the Northern Hemisphere tilted away from the sun, scientists reported with unprecedented distress that the North Polar ice cap is "falling off a cliff." One study estimated that it could be completely gone during summer in less than 22 years. Another new study, to be presented by U.S. Navy researchers later this week, warns it could happen in as little as seven years.[27]

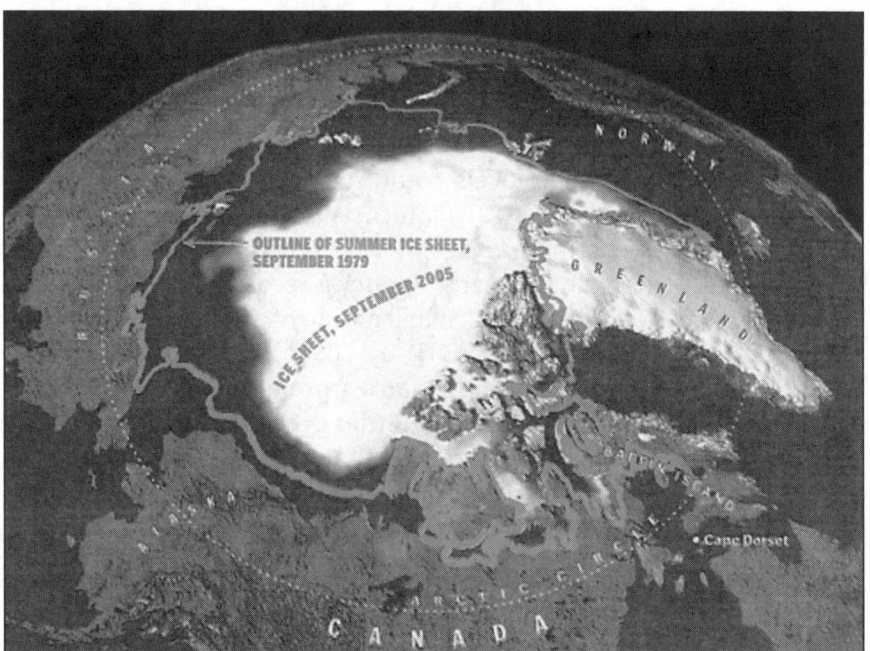

Fig. 2–4. Space photo showing the loss of Arctic ice sheet from 1979 to 2006

The IPCC identifies melting ice sheets that could lead to a rapid rise in sea levels and the extinction of large numbers of species brought about by even moderate warming of 1–3°C.[28] As global warming raises the Earth's temperature, the corresponding increase in ocean temperature provides energy for more forceful hurricanes. Recent evidence is that oceans have been warming and rising much faster than even predicted by climate change models. Decreasing moisture availability with warmer temperatures increases evapotranspiration, reduces snowpack, promotes an earlier snowmelt, and increased the U.S. fire season by 78 days over the past 20 years.[29] Global warming will reduce food production and crop yields in lower latitudes, and promote the rapid spread of infectious diseases and cardiovascular disease, while spurring competition for dwindling water resources.[30] The urgency is increasingly apparent: the IPCC Fourth Assessment Report in 2007 concluded that the evidence of human-made global warming is "unequivocal."[31]

A proverb states that "time is money." Nowhere is this more true than with international efforts to abate carbon concentrations in the atmosphere. More than a decade has transpired in the control of carbon emissions. In 1990, the world emitted about 40 billion tons of CO_2e, while today it has increased by about 40% to 55 billion tons annually. There is no sign that world carbon emissions are decreasing. Global carbon concentrations in the atmosphere are now accelerating at four times the rate they did a decade ago.

The U.S. Department of Energy forecasts that a worldwide carbon *increase* of 54% over 1990 levels could occur by 2015.[32] The Energy Information Administration (EIA) forecasts a 50% worldwide increase of carbon emissions between 2005 and 2030 as the most likely reference scenario.[33] The International Energy Agency forecasts a 25–90% increase over the same period.[34] It concluded that absent of a major policy change, CO_2 emissions could increase 130% by 2050. Most of the projected increases will occur in developing countries, projected to grow five times as fast as those from industrialized countries over the next 25 years.[37]5By 2030, developing countries are forecast to exceed CO_2 emissions from developed countries by 72%. Until 2030, worldwide use of coal is expected to increase roughly 60%.

GHGs increased about 70% between 1970 and 2004, with combustion of fossil fuels accounting for 70% of GHG emissions. Electric power generation was responsible for 40% of these CO_2 emissions, with coal-fired electric power generation accounting for about 70% of the emissions in this sector. Global energy-related emissions are expected to increase 57% from 2005 to 2030.[36] From any perspective, there is a fast-growing problem, with coal resources at the center.

According to a 2007 report from the United Nations Environment Programme (UNEP), forecasters do not see the international mix of power generation sources changing appreciably over the next several decades. Without a substantial change to renewable or other low-carbon technologies, the percentage of fossil fuels in the mix—and thus the potential sources of additional GHGs emanating from the power sector—is forecast to remain relatively constant. The International Energy Agency forecasts that by 2030, world demand for energy will grow by 59% and fossil fuel sources will still supply 82% of the total, with non-carbon renewable energy sources supplying only 6% of the total.[37]

So if this is the scientific basis, what is the role of the electric power sector in the equation? We focus there in the next chapter.

Notes

1. Pew Center for Climate Change. *Global Warming Basics.* http://www.pewclimate.org/global-warming-basics.

2. For a complete discussion of global warming, see Houghton, J. 2004. *Global Warming: The Complete Briefing.* 3rd ed. Cambridge, UK: Cambridge University Press.

3. Bryson, B. 2003. *A Short History of Nearly Everything.* New York: Broadway Books, 39.

4. Bureau of National Affairs. 2008. IPCC Official Says Industrialized Nations Must Cut Emissions up to 95 Percent. *Environment Reporter,* Sept. 26: 1,917.

5. Working Group II Contribution to the Intergovernmental Panel on Climate Change. 2007. Climate Change 2007: Impacts, Adaptation and Vulnerability. *IPCC Fourth Assessment Report.* http://www.ipcc.ch/publications_and_data/publications_and_data.htm. The IPCC Fourth Assessment Report talks of temperature increases ranging from 2.4–6.4ºC. This would yield a 7–23 inch rise in sea levels during the 21st century.

6. Hansen, Jim. 2006. The Threat to the Planet. *The New York Review of Books.* July 13, vol. 53, no. 12.

7. In 2000, anthropogenic activities emitted 320 million tons of methane and 33 million metric tons (TgN) of nitrogen oxides into the atmosphere per year. These levels are rising at a rate of about 4% per year. IPCC Working Group I. 2001. *Climate Change 2001: The Scientific Basis of Climate Change.* http://www.ipcc.ch/ipccreports/ar4-wg1.htm.

8. IPCC Working Group I. 2007. Report AR4, chap. 1: 5. http://ipcc-wg1.ucar.edu/wg1/wg1-report.html.

9. U.S. Energy Information Administration. 2005. *Emissions of Greenhouse Gases in the United States 2005: Executive Summary.* U.S. Department of Energy. 2–3. http://www.eia.doe.gov/oiaf/1605/ggrpt/summary/pdf/0573(2005)es.pdf. Frequently Asked Global Change Questions, http://cdiac.ornl.gov/faq.html.

10. Wara, Michael. 2008. Measuring the Clean Development Mechanism's Performance and Potential. *UCLA Law Review.* Vol. 55, no. 6 (August): 1,759, 1,780.

11. Hansen, James and Larissa Nazarenko. 2004. *Soot Climate Forcing Via Snow and Ice Albedos.* Proceedings of the National Academy of Sciences. January 13: 425.

12. Ramanathan, V. and G. Carmichael. 2008. Global and Regional Climate Changes Due to Black Carbon. *Nature Geoscience.* March 23: 221; Bond, T.C. and H. Sun. 2005. Can Reducing Black Carbon Emissions Counteract Global Warming? Environmental Science & Technology, July 6: 5,921.

13. Hansen and Nazarenko. *Soot Climate Forcing Via Snow and Ice Albedos,* 428.

14. U.S. Congress. House. 2007. Committee on Oversight and Government Reform 12. *Testimony of V. Ramanathan for the Hearing on Black Carbon and Climate Change,* 111th Cong., October 18: 3.

15 U.S. Congress. House. 2007. Committee on Oversight and Government Reform 12, *Testimony of Mark Z. Jacobson for the Hearing on Black Carbon and Climate Change*, 111th Cong., October 18. http://oversight.house.gov/story.asp?ID=1550; Ramanathan and Carmichael, at 226.

16 U.S. Congress. *Testimony of Mark Z. Jacobson*, 3.

17 Ramanathan and Carmichael, Global and Regional Climate Changes.

18 Ibid., 226.

19 Jacobson, Mark. 2002. Control of Fossil-Fuel Particulate Black Carbon and Organic Matter, Possibly the Most Effective Method of Slowing Global Warming. *Journal of Geophysical Research*. 107 (D19): 19. (Citing Pope III, C. A. and D. W. Dockery. 1999. Epidemiology of Particle Effects, in S. T. Holgate et al., eds., *Air Pollution and Health* 673–705, and statistics from the World Health Organization.)

20 U.S. Congress. House. 2007. Committee on Oversight and Government Reform 2-3. *Testimony of Tami Bond for the Hearing on Black Carbon and Climate Change*, 111th Cong., October 18. http://oversight.house.gov/story.asp?ID=1550.

21 U.S. Congress. *Testimony of Mark Z. Jacobson*, 5.

22 Bond, Tami. 2002. *Summary C. Aerosols*. Presented at Air Pollution as a Climate Forcing: A Workshop, Honolulu, Hawaii, April 29–May 3. http://www.giss.nasa.gov/meetings/pollution2002.

23 Ramanathan and Carmichael, Global and Regional Climate Changes, 226.

24 Ibid, 221.

25 U.S. Congress. *Testimony of Ramanathan*, 4.

26 McCarthy, Michael. Vast Ice Shelf Collapses in the Arctic. *The Independent* (London), December 30, 2006. http://www.independent.co.uk/environment/climate-change/vast-ice-shelf-collapses-in-the-arctic-430250.html.

27 Gore, Al. 2007. Speech accepting Nobel Peace Prize, Oslo, Norway, Dec. 10.

28 The IPCC 2007 report predicts that in the period 2007–2100, sea levels will rise by approximately 0.6–1.9 ft., also noting that such numbers may be increased by 3.9 to 7.8 in. if there is a continuation of rapid polar ice melt. 2007 IPCC Report on Climate Change summary available at http://www.wunderground.com/education/ipcc2007.asp.

29 See http://www.environmentaldefense.org/article.cfm?ContentID=5816.

30 Parry, M.L. and O.F. Canziani, J.P. Palutikof, P.J. van der Linden and C.E. Hanson, eds. 2007. Contribution of Working Group II to the Fourth Assessment Report of the Intergovernmental Panel on Climate Change. *Climate Change 2007*: Climate Change Impacts, Adaptation and Vulnerability. Cambridge, UK: Cambridge University Press. http://www.ipcc.ch/publications_and_data/publications_ipcc_fourth_assessment_report_wg2_report_impacts_adaptation_and_vulnerability.htm.

31 Bernstein, Lenny et al. 2007. *Climate Change 2007: Synthesis Report.* Intergovernmental Panel on Climate Change. 30. http://www.ipcc.ch/pdf/assessment-report/ar4/syr/ar4_syr.pdf.

32 Reitze, A. 2001. *Global Warming. Environmental Law Reporter.* 31. 10253, 10255.

33 Environmental Protection Agency. *The Terrestrial Carbon Cycle: Managing Forest Ecosystems.* http://www.epa.gov/wed/pages/projects/globalclimatechange/TerrestrialCarbonCycle.pdf.

34 U.S. Department of Energy. 2008. *EIA International Energy Outlook 2008.* June. www.eia.doe.gov/oia/ieo/index.html; IPCC, *Fourth Assessment Report.*

35 U.S. DoE, *EIA 2008 Energy Outlook.*

36 U.S. Government Accountability Office. 2008. *International Climate Change Programs: Lessons Learned from the European Union's Emissions Trading Scheme and the Kyoto Protocol's Clean Development Mechanism.* GAO-09-151. (November): 48.

37 International Energy Agency. 2004. *World Energy Outlook 2004.* http://www.iea.org/textbase/nppdf/free/2004/weo2004.pdf.

3) THE CRITICAL ROLE OF ELECTRIC POWER ARCHITECTURE AND CARBON

The Role of Fossil Fuels

Choice of power generation technology translates directly to the size of our carbon footprint. CO_2 is the main by-product of fossil fuel combustion, and therefore results from any energy production that uses oil, coal, natural gas, or other solid waste fuels. In 1999, all nations on Earth consumed prodigious quantities of fossil fuels: 26.7 billion barrels of oil, 81.1 trillion cubic feet of natural gas, and 2.1 billion tons of coal—all of which are decayed organic matter previously brought to life by the sun.[1] These uses of fossil fuels for energy and electricity are shown in figure 3–1. However, when burned for electric production, all release atmospheric carbon.

Fossil fuel generation results in 64% of the total atmospheric CO_2, and this amount has increased significantly since 1990.[2] Power derived from burning gaseous, liquid, and solid fossil fuels used to create electric power release copious quantities of CO_2 into the environment. Most countries are using fossil fuels, not renewable power resources, to satisfy this exponential increase in demand.

Global CO_2 emissions are rising at the rate of approximately 10% per year.[3] GHGs increased about 70% between 1970 and 2004, with combustion of fossil fuels accounting for 70% of GHG emissions, and electric power generation responsible for about 40% of these CO_2 emissions—particularly with coal-fired electric power generation accounting for about 70% of the emissions in this sector. Global energy-related emissions are expected to increase 57% from 2005 to 2030.[4]

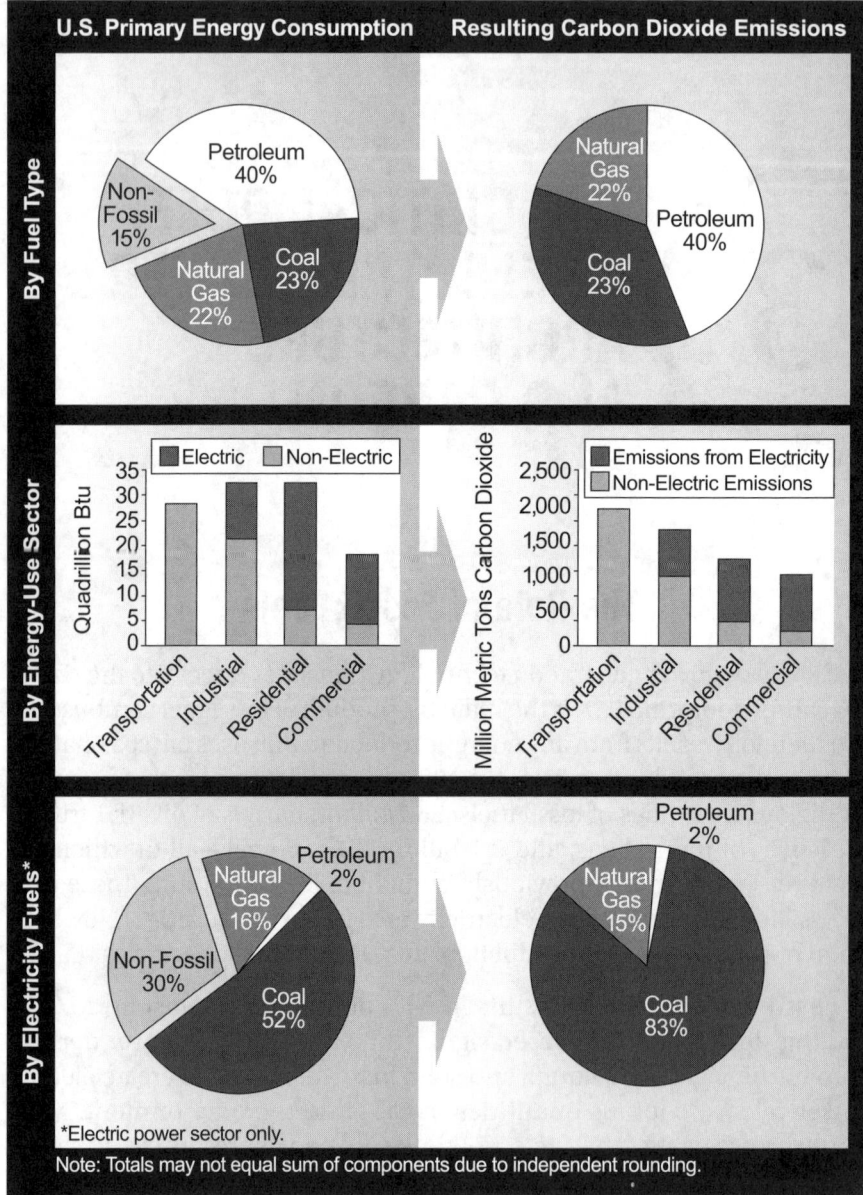

Fig. 3–1. U.S. primary energy consumption and carbon dioxide emissions in 2006 (Source: *Energy Information Administration*)

Of anthropogenic CO_2 emissions, 98% are from combustion of fossil fuels, and 83% of U.S. GHG emissions are attributed to CO_2.[5] More than one-third of CO_2 emissions are attributable to the electric power sector. Electric power is the crucial carbon-emitting sector of world economies not only because of its absolute contribution to the warming problem, but also because of its rate of growth. These emissions from stationary power production sources are increasing more quickly each year than emissions from other fossil fuel sources, including the transportation sector.

The Power Technology *FOOT* in the Carbon *FOOTPRINT*

Clearly GHGs in the 21st century are about power generation sources and means. The importance of the electric sector in global warming abatement is reflected in its changing role. In 1949, only 11% of global warming gases in the U.S. came from the electric sector; today it is more than 40%. From 1990 to 2005, 55% of the growth in CO_2 emissions occurred in the electric power sector. It is the dominant sector for recent changes in carbon emissions. The share of GHGs emitted by the electric sector in the United States are shown in figure 3–2.

The Energy Information Administration in 2008 concluded that the electric power sector offered the most cost-effective opportunities to reduce CO_2 emissions, compared to the transportation sector. So the power sector will be the carbon reduction focus, and where *additionality* has its primary application (see chapter 15). The types of technologies in the power generation capital stock largely determine the long-term concentrations of atmospheric carbon.

Energy choices and climate security are intertwined in a global helix, and must be addressed effectively at the same time. Leading climate scientists give the world less than a decade to accomplish a dramatic reversal of increasing GHG emissions, or face climate catastrophes (see chapter 4). To do this, there must be a dramatic and massive shift in the power-generating base from CO_2-emitting fossil fuels to renewable power, which has not occurred under the existing Kyoto Protocol or other policies. Major developing countries will make their investments in energy infrastructure over the next decade, which provides a particularly narrow window in which to make the best choices for their energy future.

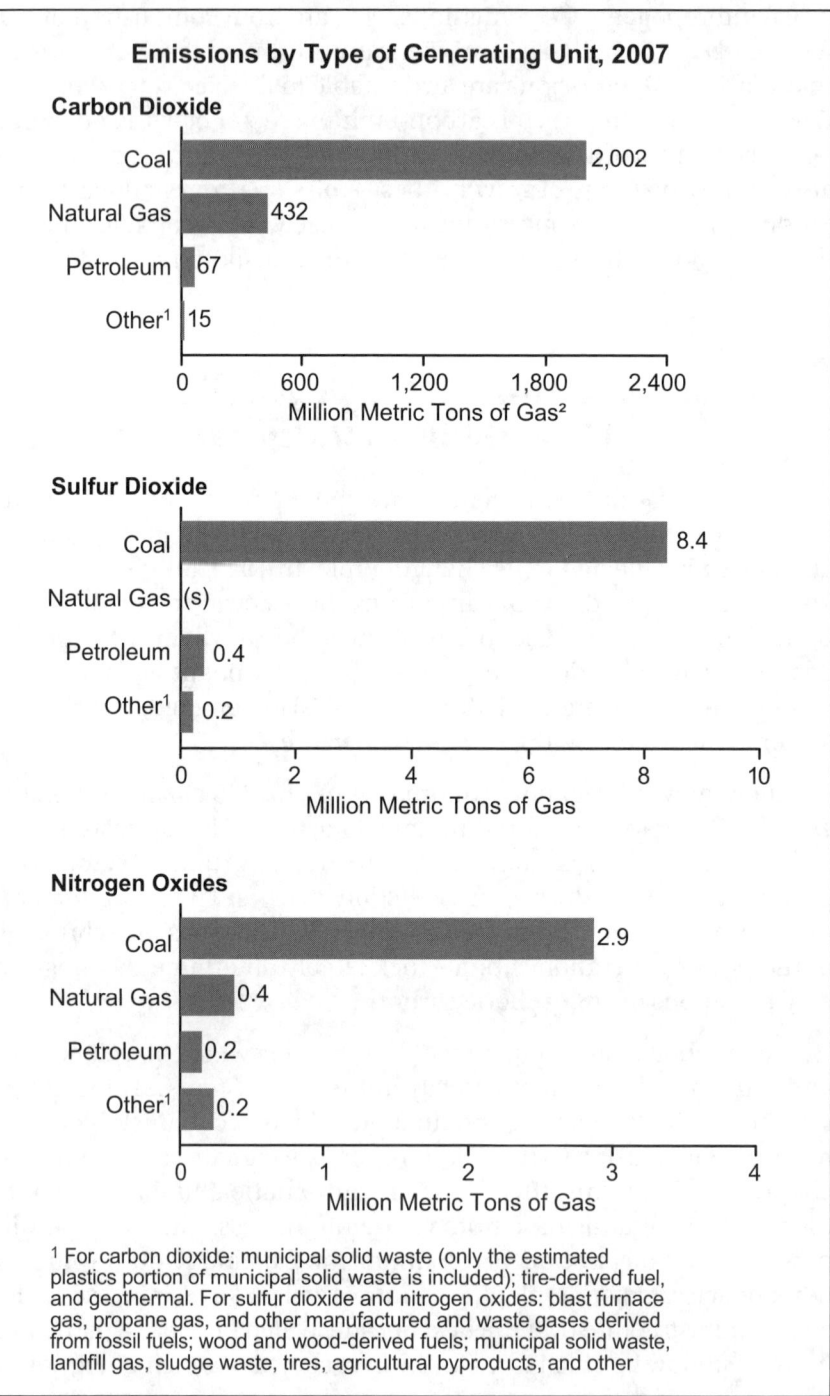

Fig. 3–2. Emissions from energy consumption for electricity generation
(Source: Energy Information Administration, Annual Energy Review, 2008)

Emissions by Sector, 1989-2007

Carbon Dioxide

(Billion Metric Tons of Gas)

Total[2]
Electric Power Sector
Industrial Sector

Sulfur Dioxide

(Million Metric Tons of Gas)

Total[2]
Electric Power Sector
Industrial Sector

Nitrogen Oxides

(Million Metric Tons of Gas)

Total[2]
Electric Power Sector
Industrial Sector

biomass; and chemicals, hydrogen, pitch, sulfur, and tar coal.
[2] Includes Commercial Sector.
(s)=Less than 0.05 million metric tons.
Sources: Tables 12.7a-12.7c.
Energy Information Administration / Annual Energy Review 2008

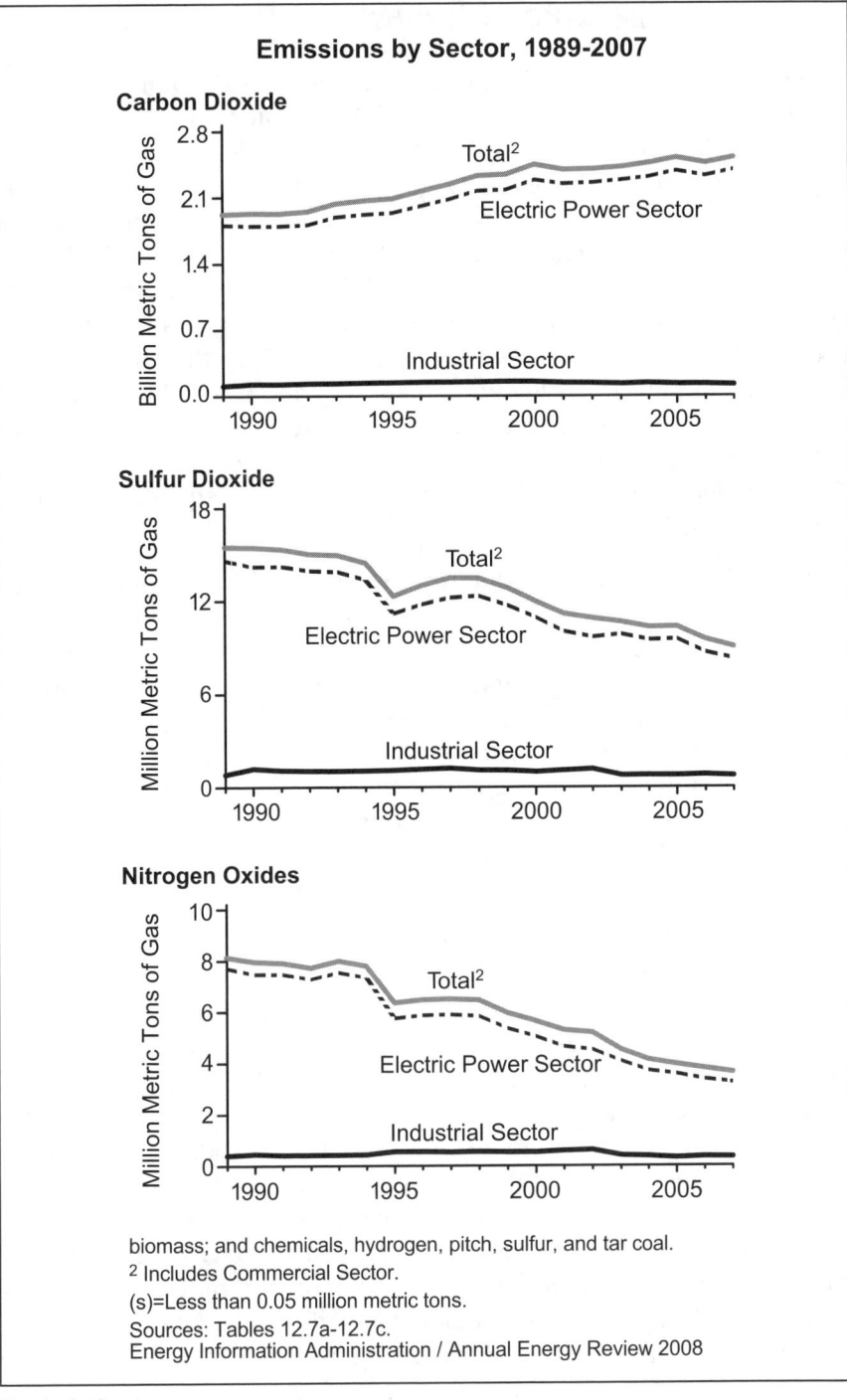

Fig. 3–2. Cont.

The Role of Renewable Power

Electric power demand is continuing to increase. At current rates of energy development, energy-related CO_2 emissions in 2050 would be 250% of their current levels under the existent pattern.[6] Unprecedented deployment of renewable energy generation alternatives will be required to alter this trend. The technology exists to accomplish this. The amount of solar radiation reflecting off the Earth is about 1,000 times the Earth's commercial energy use. In fact, no nation uses more energy than the energy content contained in the sunlight that strikes existing buildings every day.[7] The solar energy that falls on roads in the United States each year contains roughly as much energy content as all the fossil fuel consumed in the world during that same year. The estimated U.S. roof area of buildings is almost 70 billion square feet, which could accommodate 710,000 megawatts (MW) of solar photovoltaic (PV), equivalent to peak demand in the United States.

The GHG mix of electric energy sources is within legal control by government policy and incentives. Less renewable energy—by an order of magnitude compared to fossil fuels—is utilized. Total installed capacity of renewable energy power generation, excluding large hydropower, was 142 gigawatts (GW) worldwide, of which 58 GW was in developing countries (see table 3-1). While this installed capacity is but a tiny fraction of the 3,700 GW of total electricity generation capacity, installations of some technologies such as wind and solar photovoltaics are growing at over 25% per year, albeit from a small base. Still, the renewable energy used by humankind on the earth equals only about 0.01% of the total solar energy reaching the earth.[8]

These renewable resources are widely disseminated across the globe. While many nations—particularly developing nations—have no significant fossil fuel reserves of oil, coal, or natural gas, every nation has significant renewable energy in some form (e.g., hydropower, sunlight, wind, agricultural biomass waste, wood, ocean wave power, etc.). This allows for energy independence and provides a resource for domestic economic development. While the commercial and national interests involved in fossil fuel is extremely concentrated, solar energy interests and flows are much more decentralized and diverse. Whether or not governments will divert the fossil fuel vector toward renewable or other low-carbon electric energy deployment, and do so in time to avert global warming, is in part a function of whether there is a clear roadmap on how to navigate the carbon-concerned future of development.

Table 3–1. Grid-based renewable power capacity as of 2003
Source: http://www.jxj.com/magsandj/rew/2004_05/indicators_of_investment.html

Generation Type	Capacity in all countries (GW)	Capacity in developing countries (a) (GW)
Small hydro power (b)	56	33
Wind power	40	3
Biomass power (c)	35	18
Geothermal power	9	4
Solar photovoltaic power (grid-connected)	1.1	< 0.1
Solar thermal power	0.4	0
Total renewable power capacity	142	58
For comparison—large hydro power (d)	730	340
Total electric power capacity	3,700	1,300

a Developing countries are non-OECD countries plus Mexico, South Korea and Turkey, excluding countries with economies in transition. Martinot et al. (2002) included economies in transition in these totals, reflecting all countries eligible for World Bank development assistance.

b Definitions of small hydro vary by country. They usually cover hydro up to 10 MW, although this figure is up to 25 MW in India and up to 30 MW in China – thus global totals can differ greatly depending on what is counted.

c Biomass power figures exclude municipal solid waste combustion and landfill gas.

d Published hydro power figures assumed to include both large and small hydro, except in China, where these are reported separately. Total hydro is the sum of small and large hydro.

Despite the emergence of, and attention to, renewable energy sources, forecasters do not see the international mix of power generation sources changing appreciably over the next several decades. The percentage of fossil fuels in the mix—and thus the potential sources of GHGs in the power sector—is forecast to remain relatively constant. Changes in use of energy, even in modern industrial times, move slowly. Figure 3-3 sets forth the evolving pattern of energy use in the U.S. over the past four centuries. As our economy has moved from wood to coal, to oil, to gas, and perhaps in the future more to renewable energy sources, each transition reduces the amount of CO_2 released per effective unit of energy harnessed. Therefore, developing countries now choosing to deploy significant amounts of the earlier phase fossil fuels on this figure are making choices at relatively high-carbon emission parts of the curve. The International Energy Agency in Paris predicts that by 2030, world demand for energy will grow by 59% and fossil fuel sources will still supply 82% of the total, with non-carbon renewable energy sources supplying only 6%.[9]

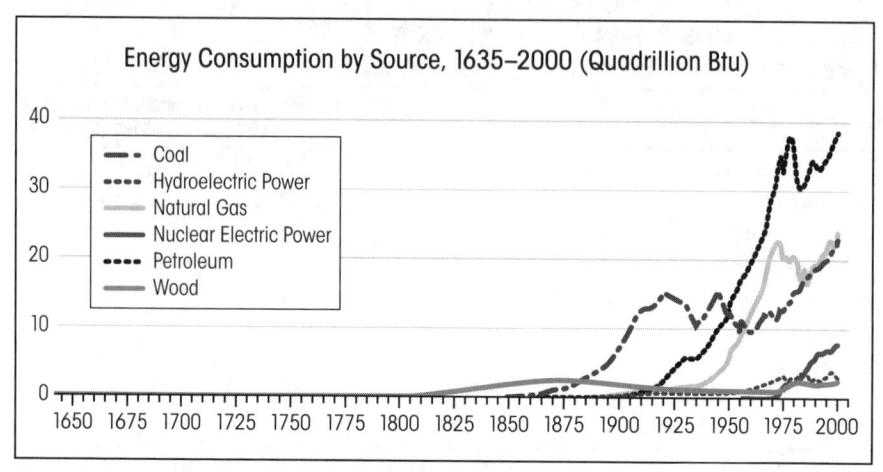

Fig. 3–3. Energy consumption in the United States since 1650

Without this roadmap, the percentage of fossil fuels in the mix—and thus the potential sources of GHGs in the power sector—is forecast to remain relatively constant. So, without a radical restructuring of the institutional framework in the power sectors of world nations, business-as-usual leads to a continuing significant increase, rather than the required decrease, in annual CO_2 emissions. This status quo approach compounds GHG concentrations in the atmosphere, and the repercussions of rapid global warming.

The Key Role of Developing Countries in Future Power

More than 80% of world nations would be characterized as developing, rather than developed. The average annual growth rate in primary energy use in developing countries from 1990 to 2001 grew by 3.2% per year, compared in industrialized countries where growth over the same period was 1.5% annually.[10] As much as a 4% per year increase in demand by developing countries over the next 20 years is predicted by the International Energy Agency.[11] It forecasts that two-thirds of all future energy demand will emanate from just China and India. The U.S. Department of Energy forecasts that energy demand in developing Asia will double over the next 25 years.

This technology choice is critical in developing nations. Although the majority of anthropogenic CO_2 emissions is generated in developed countries, this balance is shifting toward developing countries. World CO_2 emissions, currently estimated at about 25 gigatons (Gt) annually, are slightly dominated by Organization for Economic Cooperation and Development (OECD) nations compared to developing countries.[12] The crossover point is projected to be in 2020 when OECD countries and developing countries each emit roughly comparable amounts of CO_2 into the atmosphere. By 2030, the position of developed and developing nations will have reversed, with developing countries providing the dominant share of CO_2 emissions, and increasing over time into the future.

Focus on Asia's Rapid Development

The trend is important. Developing nations are expected to emit a majority of CO_2 emissions before 2035, while China recently surpassed the United States as the largest CO_2 emitter in the world. China already has the highest emissions in the world per unit of gross national product (GNP) by a factor more than double other nations. Thus the critical reality: developing nations represent the geographic core of exponential increase in CO_2 emissions, and are where global warming law and policy must focus. To ignore them, as the Kyoto Protocol does, in the world's carbon regulatory regime, is to invite broad policy failure.

GHG emissions among developing nations feature Asia as the proverbial 800-lb. gorilla. China and India harbor around one-quarter of the world's coal reserves, and are deploying them rapidly to fire electric power plants. China currently meets 70% of its electricity demand through coal plants, the most prolific emitters among fossil fuel plants in terms of both CO_2 and particulate matter; 57% of India's electricity comes from coal. India has targeted 100,000 MW in new capacity over the next 10 years.[13] Even Vietnam is planning on adding scores of additional new hydroelectric and oil-fired plants by 2010.

The pace and velocity of electrification and power capacity addition in developing countries are staggering. Specifically, to meet its growing energy needs, China is planning on rolling out 100 new large power plants by 2020, including nuclear, hydropower, and, principally, coal-fired plants. China is adding 1 GW of new coal-fired power every week, which short-term is on a 700% faster pace than the plan for 100 plants over a 13-year scenario.[14] Fossil fuel-fired plants contribute to climate change. Predictions are that by the year 2030, coal-fired power in India and China will add 3,000 million extra tons of CO_2 to the atmosphere every year.[15] Some projections estimate that by 2030, China's GHG emissions will quadruple and Asia alone will emit 60% of the world's carbon emissions.[16] Therefore, just these new CO_2 emissions from China and India electric power sectors will constitute approximately 10% of all world CO_2 emissions from all sources.

Before recently surpassing the United States for gross carbon emissions, Chinese total installed electric generation capacity grew from 65 GW in 1980 to 353 GW in 2002, making it the largest carbon emitter in the world. Chinese electricity demand between 1996 and 2000 grew at an average of 6.3% annually, and is expected to almost match this pace into the future. In order to avoid shortages and satisfy demand, China would have to increase electric capacity by approximately 40 GW annually.

Similar patterns of rapid electrification utilizing fossil fuels are evident in other major developing nations. By 2025, one quarter of the world's population will be living in Asian cities. India, with 1.1 billion people, is the second most populous nation, on its way to being the first most populous nation, and contains the world's fourth largest coal reserves. India maintains about 123,000 MW of generation capacity, compared to approximately 400,000 MW in China and approximately 1,000,000 MW in the United States.

Urbanization and population growth in India have driven a 208% growth in India's energy consumption in the last 20 years; India has targeted 100,000 MW in new power generation capacity over the next 10 years.[17] In June 2008, just before the G8 conference, India rejected any commitment to mandatory GHG cuts. With China and India building almost a new coal plant each week, just these *new* CO_2 emissions from their electric power sectors will constitute approximately 10% of all world CO_2 emissions from all sources.

Actually, none of the countries with the largest coal reserves—United States, China, India, Indonesia—has a carbon policy to regulate the release of CO_2 from the deployment of such coal reserves. None had joined or is regulated by the Kyoto Protocol as of 2009. It has been predicted that Russia's CO_2 emissions will climb dramatically by 2020, due to a 123% increase in Russian coal consumption.[18] Italy over the next five years is expected to more than double the percentage of coal used to generate electricity.[19] Germany, England, and other major countries also are planning new coal plants.

Coal and Renewable Power

According to climate scientist James Hansen, "Building a new coal-fired power plant is ill conceived...we need a moratorium on coal now, with phase-out of existing plants over the next two decades."[20] According to former vice president Al Gore, "We also need a moratorium on the construction of any new generating facility that burns coal without the capacity to safely trap and store carbon dioxide."[21] However, coal remains attractive to power suppliers because of abundance, its dispersion as an energy resource across mineral seams in many countries, and the fact that there is no coal cartel, as there is with oil producers.

According to former British prime minster Tony Blair, "the vast majority of new power stations in China and India will be coal-fired; not 'may be coal-fired', will be."[22] China's future energy use is projected to grow faster than its GNP.[23] Over two-thirds of its energy is produced from coal. In 2007, China built more new coal-fired power plants than Britain, the seat of the coal-fired industrial revolution, built in its entire history.

If China used energy as America does, world energy consumption would double, requiring five more Saudia Arabias of oil consumption to meet this demand.[24] China's energy demand has increased faster than its economy over the past decade. Each year, China adds 40 times more new coal capacity than new wind power. China now contains 7 of the 10 most polluted cities in the world. According to Hansen, "The stakes, for all life on the planet, surpass those of any previous crisis....If we stay our present course, using fossil fuels to feed a growing appetite for energy-intensive lifestyles, we will soon leave the climate of the Holocene, the world of prior human history."[25]

GHGs in the 21st century are about power generation. The single-point nature of power plants' emissions, the centralized nature of most power plant decisions in developing nations, and the exploding demand for electricity, make electricity generating plants the logical choice for the intelligent legal and policy assault on GHG emissions. There are solutions with renewable electric power generation technologies. Renewable energy technologies in many instances are justified without regard to their GHG benefits in developing nations. This is particularly true where new electric supply infrastructure is being created or extended. These renewable technologies can be installed off-grid to provide regional electricity supply, thus negating the need to extend the grid wires to remote locations. This conserves materials, resources, and potential line losses during long-distance transmission of centrally produced electricity to remote locations.

The funding now of renewable energy projects worldwide, and especially in fast-growing and industrializing Asia, is necessary to prevent these nations from becoming even more reliant on a high-carbon fossil fuel-based generation infrastructure (see chapter 19). At the Bonn International Conference on Renewable Energies in June 2004, China committed to adding about 6 GW of renewable energy annually until 2020. The International Energy Agency projected that it will require an investment of $16 trillion by 2030 to meet the world's energy requirements, with $5 trillion of that amount allocated to electric power production, primarily in Asia and Africa.[26] In the past 10 years of Kyoto's attack on carbon, this increasing energy demand has progressed much more vigorously than the transition to renewable sources of power generation.

The pressing issue is how much time do we have? The next chapter addresses whether the world is immediately approaching a *tipping point* at which point there is looming catastrophe.

Notes

1. See, *National Energy Foundation's Fuel Consumption Statistics* available at http://www.nef1.org/ea/eastats.html.

2. Energy Information Administration. 2005. *Emissions of Greenhouse Gases in the United States 2005: Executive Summary*. U.S. Department of Energy, 2–3. http://www.eia.doe.gov/oiaf/1605/ggrpt/summary/pdf/0573(2005)es.pdf; Carbon Dioxide Information Analysis Center. *Frequently Asked Global Change Questions*. http://cdiac.ornl.gov/faq.html.

3. See, Purdy, Ray. 2006. The Legal Implications of Carbon Capture and Storage under the Sea. *Sustainable Development Law & Policy*. American University College of Law. (Fall): 22.

4. U.S. Government Accountability Office. 2008. *International Climate Change Programs: Lessons Learned from the European Union's Emissions Trading Scheme and the Kyoto Protocol's Clean Development Mechanism*. GAO-09-151. (November): 48.

5. U.S. Department of Energy. 1999. *EIA Emission of Greenhouse Gases in the United States, 1998*. http://www.eia.doe.gov/emev/plugs/plgg98.html.

6. International Energy Agency. 2006. *Energy-Technology Perspectives: Scenarios and Strategies to 2050*. http://www.iea.org/speech/2006/ramsay/etp-beijing.pdf.

7. Ferrey, Steven. 2007. *Environmental Law: Examples & Explanations*. 5th ed. New York: Aspen Publishers. 526.

8. Ferrey, Steven. 2009. *The Law of Independent Power*. Eagan, Minnesota: Thomson/West Publishing. Sec. 2:1.

9. International Energy Agency. 2004. *World Energy Outlook 2004*. http://www.iea.org/textbase/nppdf/free/2004/weo2004.pdf.

10. Ibid., 31.

11. International Energy Agency. 2002. *World Energy Outlook 2002*. http://www.iea.org/textbase/appdf/free/2000/weo2002.pdf.

12. See, Purdy, *The Legal Implications of Carbon Capture and Storage under the Sea*, 23, table 1. OECD and developing countries collectively constitute more than 90% of all CO_2 emissions and are projected to continue this percentage over time.

13. U.S. Department of Energy. 2009. *India Country Analysis Brief*. http://www.eia.doe.gov/cabs/india.html. March.

14. See, Purdy, *The Legal Implications of Carbon Capture and Storage under the Sea*, 23.

15. UK House of Commons. 2006. Science and Technology Committee. *Meeting UK Energy and Climate Needs: Role of Carbon Capture and Storage*. First report of Session 2005–06. Vol. 1, HC-578-1.

16. Cooper, Deborah. 1999. The Kyoto Protocol and China: Global Warming's Sleeping Giant. *Georgetown International Environmental Law Review*. 11: 401, 405.

17. U.S. Department of Energy. 2009. *India Country Analysis Brief*.

18. Carlsen, Paul. 2008. Clean Coal 'Optimists' Win Show-of-hands Votes at EEI London. *Platts Electric Utility Week*. (March 17): 1, 15.

19. Rosenthal, Elisabeth. 2008. Europe Turns to Coal Again, Raising Alarms on Climate Change. *New York Times*. April 23, A1.

20. Ibid.

21. Gore, Al. 2007. Speech accepting Nobel Peace Prize, Oslo, Norway, Dec. 10.

22. Blair, Tony. 2008. *Breaking the Climate Deadlock: A Global Deal for Our Low-Carbon Future*. The Climate Group. 6.

23. Borgford-Parnell, Nathan. 2008. China's Renewable Energy Law. *Sustainable Development Law & Policy*. American University, Washington College of Law (Winter): 45.

24. Chandler, William. 2008. Breaking the Suicide Pact: U.S.-China Cooperation on Climate Change. *Carnegie Endowment for International Peace*. Policy Brief 57, (March): 1, 6.

25. Hansen, James et al. 2008. *Target Atmospheric CO_2: Where Should Humanity Aim?* Working Paper.

26. International Energy Agency. 2003. *World Energy Investment Outlook 2003*.

4) THE TIPPING POINT: TIME AS THE ENEMY

"We have passed that and some other tipping points in the way that I will define them . . . We have not passed a point of no return. We can still roll things back in time—but it is going to require a quick turn in direction."[1]

The Critical Point

In 2009, the United Nations Environment Programme forecast urgency of coming "tipping points . . . that will alter regional and global environmental balances . . . irreversible within the time span of our current civilization." Biologic carbon sinks, such as forest and ocean carbon uptake, are showing signs of increased stress under rising temperatures. A recent assessment is that we need to limit the increase in Earth "surface temperature to no more than 2-2.5 degrees Centigrade above . . . the 15 degree Centigrade" Earth temperature present at the time of the American Revolution, to avoid catastrophic effects of global warming.[2] This will require a sharp reduction of emissions over the next generation, and to near zero by 2100. This will only be possible if we can demonstrate that a modern society can function without reliance on technologies that release carbon dioxide. Recent CO_2 levels in Hawaii, isolated from other landmasses, are shown in figure 4-1.

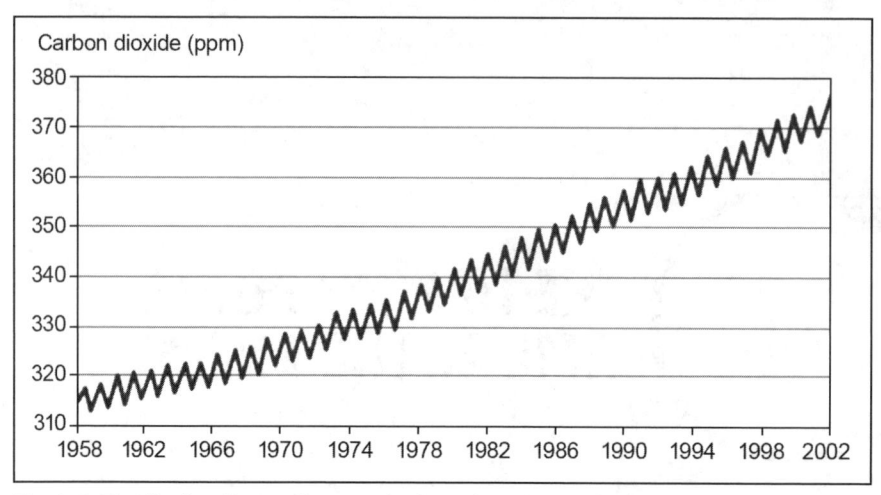

Fig. 4–1. The Keeling Curve of atmospheric carbon dioxide from Mauna Loa, Hawaii

The EU carbon reduction program is now in its fifth year, and has seen CO_2 emissions in the European Union increasing each year. EU levels have increased faster than increases in the United States during 2007, which was not regulating carbon. The World Wildlife Fund accused the European Union of "playing tricks on the atmosphere" by claiming to achieve a 20% reduction by 2020, when actually achieving in-country only about one-quarter of the 20% goal.

Several leading carbon scientists, including leaders at NASA and the Obama administration's top science advisor, warn that we have until 2015, as a world economy, to drastically reduce carbon emissions or risk catastrophe.[3] But not only is it unlikely that carbon will have been dramatically reduced worldwide by 2015, it is unknown whether it will have been reduced at all, or if that will be enough. Population growth, alone, threatens achievement of reduction of world CO_2 emissions.

In fact, the UN IPCC Fourth Assessment Report in 2007 may already be eclipsed by events.[4] That report did not take account of post-2000 increase in emissions in major Asian countries or the increasing ice melt in Greenland and West Antarctica, which accelerate impacts.[5] Rather than allowing a small atmospheric concentration until emissions plateau and decrease, it may require an immediate decrease. Both financially and in terms of moving developed and developing nations of the world alike, it is estimated to require an investment of $200 billion over the next two decades, just to hold world carbon emissions at *current* levels, let alone reduce them.

Moreover, at the pace that things aren't moving, it quickly can become too late. Global CO_2 atmospheric emission concentrations during the past three centuries are displayed in figure 4-2. NASA climatologist James Hansen notes that merely waiting less than a decade, until 2018, to stop the "growth of greenhouse gas emissions," then we reduce the probability to near no chance to avoid catastrophic effects of warming.[6] Hansen forecast this to exceed the tipping point once the atmosphere exceeds 400-425 ppm. At 450 ppm, there is no ice left on the planet according to Hansen.[6][7] In 2006, Hansen stated that "[W]e have at most ten years—not ten years to decide upon action, but ten years to alter fundamentally the trajectory of global greenhouse emissions."[8]

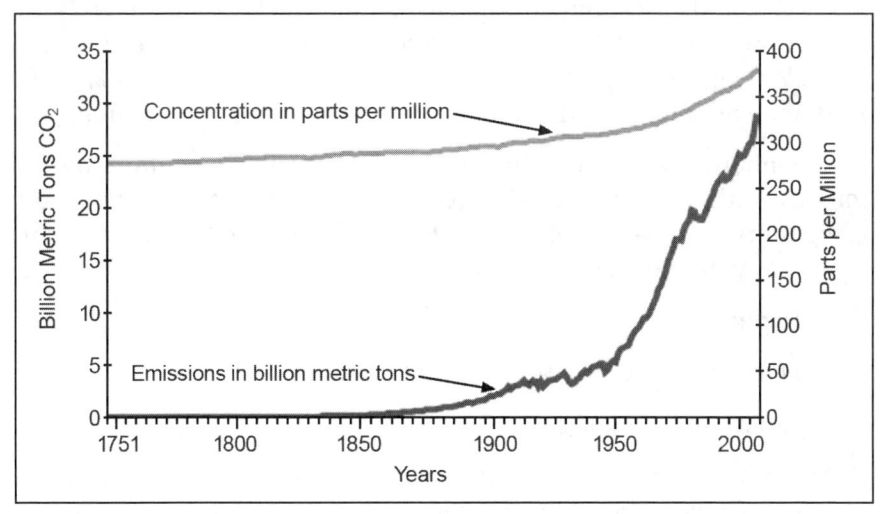

Fig. 4-2. Carbon dioxide emissions and carbon dioxide concentrations 1751-2004 (Source: Oak Ridge National Laboratory)

Since the beginning of the Industrial Revolution, CO_2 has increased by about one-third, to 382 ppm. Recent studies have forecast that the world, even with the Kyoto Protocol, is likely to reach the maximum amount of annual emissions within 15 years to hold the limit of climate increase to 2°C. To hold to this limit and given the 100 years that CO_2 emissions remain in the atmosphere, it is estimated that only about one-third of the known reserves of fossil fuels will be able to be burned.

A top official with the IPCC has indicated that developed nations will need to slash CO_2 emissions by 80-95%—almost entirely—by 2050 to hold GHGs to 450 ppm in the atmosphere. According to John Holdren, the current White House science advisor, if U.S. greenhouse emissions plateau even in five years in 2015, we already have reduced our chances by 50% to avoid climate catastrophes.[9] Time may be of the essence. Certainly, time is money. The world to date seems to be seizing neither.

The Linkages and Response

To alter this trajectory, it will require strong international agency involvement. Unfortunately, holding the current line of increases is not a satisfactory event. There must be a significant immediate reduction in world carbon emissions and concentrations in the atmosphere to mitigate global warming, according to the scientific consensus. Because world CO_2 and GHG emissions persist for a century (see chapter 2), we are now reaping the GHG effects emitted during World War I. However, World War I is not the problem, as much as the past half century of prolific emission and lingering GHG atmospheric emission concentrations.

Greenhouse gases are formed naturally, as well as by anthropogenic sources. Anthropogenic emission of greenhouse gases is driven by a fairly straightforward relationship, where GHGs are a function of a three-part algorithm with the following inputs:

- Population
- Degree of development and electrification
- Choice of technology

Affecting any of the three elements in this equation changes the emission of atmospheric gases that drive the models of global warming. To mitigate global warming, we must immediately influence some or all of these three variables. However, two of these variables seem beyond the reach of most governments; only one is within the control of world environmental regulatory institutions.

Population

First, there is a direct, if complex, link between population and GHG emissions. There is little doubt that world population will increase significantly during the next 50 years, especially in less industrialized, poorer developing nations. World population could reach 8 billion people by 2020 and 9–10 billion by 2050. East Asia (including China) and South Asia now contain more than 3 billion of the world's 6 billion population. By 2025, one-quarter of the world's population will be living in Asian cities.

There are, however, still between 1.6 billion and 2 billion people worldwide with no access to modern energy services or electricity and 56% of the world's rural population lacks access to energy services.[10] Due to population growth, the number of people without access to modern energy services is increasing at approximately 30 million people per year.[11] Most of these people will desire, and ultimately receive, electricity service at some future time. Population is not a variable that is controllable with the governmental systems employed in most of the countries of the world or by international organizations.

Degree of development and electrification

The second input in the algorithm of carbon, the degree and energy intensity of development, can be influenced by regulatory policy, but it is not clear that governments of the world are willing to attempt to control this variable. Migration of rural populations to cities and urbanization will dramatically increase per capita electric consumption in developing nations. Developing nations view electrification and a higher carbon economy to be the signature of progress and development. Direct benefits from electric energy access include reliable lighting, heat and refrigeration, health benefits due to enhanced indoor air quality from the cessation of use of polluting fuel sources, reduced fire hazards, higher quality health services equipment, greater business productivity, and increased opportunities for education.

The fundamental touchstone technology for development in all nations is electricity. Electricity has been described as an agent of technological progress. As electricity is used in place of fossil fuels and human labor, less overall energy is used and more productive and efficient operations occur in certain segments of society. Electricity is so indispensable that it has become "transparent to most users, at least until there is an outage."[12]

Statistics of energy use illustrate this intensification phenomenon. The average annual growth rate in primary energy use in developing countries from 1990 to 2001 grew by 3.2% per year, compared in industrialized countries where growth over the same period was 1.5% annually.[13] As discussed in chapter 3, as much as a 4% per year increase in demand by developing countries over the next 20 years is predicted by the International Energy Agency. The U.S. Department of Energy forecasts that energy demand in developing Asia will double over the next 25 years.[14] The International Energy Agency in Paris forecasts that two-thirds of all future energy demand increases will emanate from just China and India.

The energy needs of countries outside the OECD will require an investment of some $2 trillion to install approximately 1,900 gigawatts of new electric generating capacity by 2025.[15] The International Energy Agency projected that it will require an investment of $16 trillion by 2030 to meet the world's energy requirements, with $5 trillion of that amount allocated to electric power production, primarily in Asia, Latin America, and Africa.[16] It is expected that global energy use will double by 2040 and triple by 2060, creating a tremendous demand on existing fuel sources.

In a world where burning fossil fuels is the dominant electric energy signature, intensified use of electric power foretells a direct increase in carbon emissions. Some projections estimate that by 2030, China's GHG emissions will quadruple and Asia alone will emit 60% of the world's carbon emissions.[17] To cope with the increased electrification that accompanies the substantial increase in per capita energy use that will occur in developing nations in the next decades, the world may have to achieve a reduction of CO_2 of up to 50% during the 21st century. While increased energy intensity in developing nations may be difficult to control, the future is directly dependent on whether fossil fuels or renewable technologies are chosen now to generate power to meet this new, more intensive electricity demand.

Choice of technology

Thus, the third and the only one of the inputs in the global warming equation that can be influenced dramatically now by policymakers is the choice of technology for electrification and development. There is a policy choice involved between conventional and alternative resources. It is not an either-or choice. There is no small choice of energy infrastructure in shaping the carbon intensity of power production. The balance chosen between conventional and alternative electric resources has immense implications for the emission of greenhouse gases.

The choices for many developing countries are challenging. We stand at a crossroad in time because in the next two decades, there will be a massive electrification of developing nations. During the next decade, developing nations are choosing whether to deploy conventional fossil-fired or sustainable renewable options to generate electricity. Once installed, those facilities will remain in place, contributing to global warming or not, often for 40 years and in many cases longer. These choices in energy technology made now certainly will be the signature of our carbon footprint during the crucial period of the next half-century, during which we may pass the point of no return in terms of global warming.

Atmospheric CO_2 levels now are approximately 33% higher than in preindustrial times. By 2100, IPCC models project the average global surface temperature to warm anywhere from 1.4 to 5.8°C.[18] The risks of climate change include the potential for large-scale and possibly irreversible impacts on continental and global scales. The critical path timing of these decisions is now.

Having surveyed the science of global warming, the role of electricity in climate change, and the tipping point of critical timing issues, the next question is how the world and regions of the world are regulating climate change. We turn now to the next section that explores this issue of existing regulatory mechanisms in place or enacted to control global warming emissions.

Notes

1. Borenstein, Seth. 2007. Ominous Arctic Melt Worries Experts. *Associated Press*. Dec. 12.

2. Air and Waste Management Association. 2008. Prospects for Future Climate Change and the Reasons for Early Action; See also, Blair, Tony. 2008. *Breaking the Climate Deadlock: A Global Deal for Our Low-Carbon Future*. The Climate Group. 9.

3. See, McKibben, Bill. 2006. How Close to Catastrophe? *New York Review of Books*. Nov. 16, 23. Book review cites climatologist Jim Hansen to the effect that we have only until 2015 to reverse carbon emissions or face radically changing the planet; Hansen, James. 2006. The Threat to the Planet. *New York Review of Books*. July 13.

4. Working Group I Contribution to the Intergovernmental Panel on Climate Change. 2007. *Climate Change 2007: Synthesis Report*. http://www.ipcc.ch/publications_and_data/publications_and_data.htm.

5. See, Revkin, Andrew. 2008. Seasonal Factor Seen in Melting and Ice Shifts in Greenland. *New York Times*. July 4: A8. This article discusses melting of Greenland ice sheet.

6. Chase, Robin. 2008. Get Real on Global Warming Goals. *Boston Globe*. April 22: A15.

7. For Hansen's paper on tipping points, see Hansen, James et. al. *Target Atmospheric CO_2: Where Should Humanity Aim?* http://www.columbia.edu/~jeh1/2008/TargetCO2_20080407.pdf. "If humanity wishes to preserve a planet similar to that on which civilization developed, paleoclimate evidence and ongoing climate change suggest that CO_2 will need to be reduced from its current 385 ppm to at most 350 ppm . . . If the present overshoot of this target CO_2 is not brief, there is a possibility of seeding irreversible catastrophic effects." For a more general op-ed on the subject, see McKibben, Bill. 2008. Civilization's Last Chance: The Planet Is Nearing a Tipping Point on Climate Change, and It Gets Much Worse Fast. *Los Angeles Times*. May 11. http://www.latimes.com/news/opinion/la-op-mckibben11-2008may11,0,7434369.story.

8 Hansen. The Threat to the Planet.

9 Chase, Robin. 2008. Get Real on Global Warming Goals. *Boston Globe*. April 22: A15.

10 Dubash, Navroz, ed. 2002. *Power Politics: Equity and Environment in Electricity Reform*. World Resources Institute, sec. 1. http://archive.wri.org/publication_detail.cfm?pubid=3159#1.

11 Goldemberg, Jose and Thomas Johansson, eds. 2004. *World Energy Assessment: Overview 2004 Update*. United Nations Development Programme World Energy Council.

12 See introduction of, Gellings, Clark and Richard Lordan. 2004. The Power Delivery System of the Future. *Electricity Daily*. (January–February).

13 Goldemberg and Johansson. *2004 World Energy Assessment*, 31.

14 See *International Energy Outlook, supra*.

15 Fritsche, Uwe and Felix Chr. Matthes. *Changing Course: A Contribution to a Global Energy Strategy*, Heinrich Boll Foundation, Paper No. 22, prepared for the World Summit 2002.

16 International Energy Agency. 2003. *World Energy Investment Outlook 2003*. (November): 3, 343.

17 See Cooper, Deborah. 1999. The Kyoto Protocol and China: Global Warming's Sleeping Giant. *Georgetown International Environmental Law Review*. 11: 401, 405.

18 Working Group II Contribution to the Intergovernmental Panel on Climate Change. 2007. Climate Change 2007: Climate Change Impacts, Adaptation and Vulnerability. *IPCC Fourth Assessment Report*. http://www.ipcc.ch/publications_and_data/publications_and_data.htm.

PART II

CAP-AND-TRADE CARBON REGULATORY MECHANISMS IN PLACE ACROSS THE WORLD

5) THE KYOTO PROTOCOL: THE WORLD CARBON MODEL

The Basic Kyoto Architecture

The Kyoto Protocol, not ratified prominently by the United States and, for a time, Australia, requires 38 developed nations to reduce CO_2 emissions an average of 7% below 1990 baseline levels by 2012. The other GHGs must be reduced 5-7% below either their 1990 or 1995 baseline levels by 2008 to 2012. The Kyoto Protocol in 1997 assigned to each Annex I country (those developed nations whose carbon emissions would be regulated) a maximum quantity of allowed GHG emissions for the period 2008-2012. Developing nations successfully resisted efforts to include them in binding international obligations and opposed encouraging their voluntary commitments to GHG reduction.

Under Kyoto there is no responsibility assigned to developing countries. Kyoto reflects "common but differentiated" responsibilities between developed and developing countries.[1] The largest CO_2 emitter in the world, China, is not covered as an Annex I country. Nor are any of the countries of the world with the five largest populations.

The Kyoto Protocol is a cap-and-trade regulatory scheme. Each of the 38 regulated developed nations is allocated a national emissions cap, which applies to certain large industrial emitters of carbon within the country. At the end of each compliance period (year), each emitter must have acquired, through allocation from its government or by purchase or trade, enough credits to cover its emissions of carbon during that period. In essence, each emitter must cover its emissions with regulatory allowances or newly created and/or traded offset credits to emit carbon.

Annex I regulated countries must set up national registries to issue their internationally assigned amount units (AAUs).[2] Registry removal units (RMUs), reflecting removal of GHGs due to forestry and land-use practices, also are tracked.[3] Each AAU and RMU is tracked with a unique serial number. AAUs and RMUs are converted into emission reduction units (ERUs) for international trading purposes. Emission trading is allowed under the Kyoto Protocol. Any parties can purchase credits, even if they do not themselves require them for compliance. This includes those traders who wish to speculate in these regulatory commodities.

Kyoto Offsets

The Clean Development Mechanism

Under its Clean Development Mechanism (CDM), Kyoto includes the creation of *offsets*, called certified emission reductions (CERs). Offsets are carbon reductions in unregulated projects that can be used to create a credit for that reduction as an allowance for greater carbon emissions by the eventual holder of the credit. Including offsets in a cap-and-trade system offers several advantages:

- It allows lower-cost reduction opportunities outside the capped countries to be pursued as lower-cost reduction options.
- Economic sectors that are covered by the carbon emissions caps can be the source for reductions. This can include emission sources not otherwise cost-effectively addressed.
- They can promote technology transfer to developing countries.

Industrial emitters in each country are able to trade emission credits or create new credits through mechanisms to possess additional credits. The Clean Development Mechanism allows projects that reduce greenhouses gases in developing nations to earn certified emission reductions for each ton of CO_2-equivalent of GHG reduced.[4] Those CERs are then traded or sold to activities in Annex I developed countries, which increases that country's emission cap allocated in the protocol. All emissions reduction CERs certified under the CDM are required by the protocol to be voluntary, real, and additional to any that would occur in the absence of the CDM credit system.[5]

The CDM apparatus emerged as a last-minute compromise creation at the 1997 Kyoto Conference. It is patterned on the U.S. sulfur dioxide (SO_2) trading experience. The requirement for CDM CERs also includes the certification by the host developing nation that the project supports its goals for sustainable development. Sustainable development has been defined as "development that meets the needs of the present without compromising the ability of future generations to meet their own needs."[6] Long-term renewable energy developments clearly satisfy this definition, while many of the other CDM projects that have created CERs may be more questionable. CERs (other than for afforestation) have a 7-year lifetime, with the possibility of two renewals, for a total of 21 years, or, in the alternative, one 10-year lifetime. Some CERs related to forestry projects are deemed temporary for a period up to 60 years, subject to verification on a recurring 5-year basis that burning or logging do not later release carbon from the forest.[7]

CDM projects must be approved by regulatory agencies under the Kyoto Protocol established by the host country. These agencies are known as Designated National Authorities (DNAs). CDM registered projects by host country DNAs are shown in figure 5–1. The Kyoto Protocol process

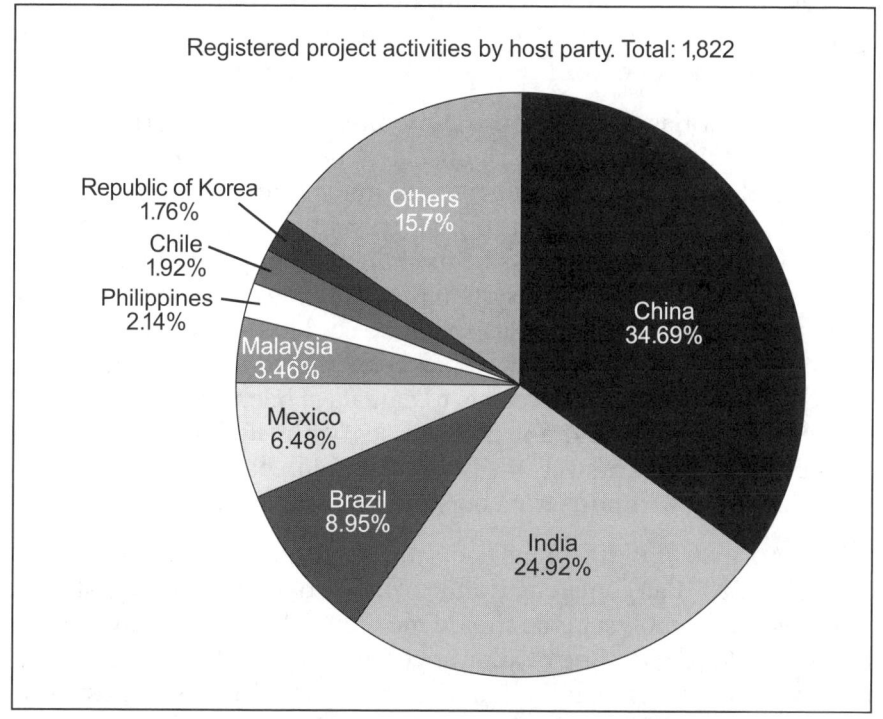

Fig. 5–1. Registered CDM projects with Designated National Authorities by country

requires 18–24 months to register and verify CERs. It is estimated that the cost of developing a new methodology for approval of CDM projects is approximately $150,000.[8] Methodologies often require an average of 280 days for approval.

CDM projects dominate the action compared to joint implementation (JI) projects (discussed later in this chapter) under the Kyoto Protocol. The volume of CDM CERs created was approximately 40 times that of joint implementation ERUs in 2006. There are almost a thousand CDM projects, with twice that many in the project development pipeline. The existing CDM projects have generated 117 million issued CERs, with an estimated 2.6 billion CERs to be generated by 2012. This would represent the equivalent in offsets equal to almost 10% of monitored CO_2 emissions.

The future of the Kyoto CDM Program

The number of projects in the CDM pipeline represents almost 3% of Annex B country 1990 GHG emissions for each year of the first commitment period. By 2012, the CDM mechanism will have produced enough carbon offsets to equal the carbon emissions of the United Kingdom over three years. The value of carbon credits and offsets is forecast to increase from €8.5 billion currently to €200 billion. Each CER generated in a developing country increases the GHG emissions allowed to be emitted in an Annex I country by the eventual owner of the CER.

The media has questioned the credibility of CDM carbon offset projects and the efficacy of such offsets.[9] A developing world producer of HCFC-22 can earn more than twice as much from its CDM subsidy as it can gross from the sale of its primary product, tripling revenues and profits (see chapter 2).[10] It has been charged that it is quite likely that the sector is also gaming the system by modifying its behavior in order to generate extra credits that can then be sold in developed countries for compliance obligations. What has been reaped by the CDM part of the Kyoto Protocol? CDM projects to date have been limited to a small number of countries and only a few gases, with "little contribution to sustainable development."[11] The European Union has announced that it may reduce CDM imports from outside the European Union after 2012.

The April 2008 Bangkok talks on the Kyoto Protocol, following the December 2007 Bali conference round of talks, concluded that a post-2012 international carbon scheme should look much like the pre-2012 Kyoto regime, including trading of allowances and the creation of additional credits or offsets through the JI and CDM mechanisms. The most cost-effective solution in developing countries, including forest preservation (see chapter 20), would thus be off the Kyoto table. Enhancements of these capabilities will require aid for viable international frameworks.

Joint implementation as offset credit transfer

A second mechanism under the Kyoto Protocol for compliance is joint implementation (JI), where developed nation signatory parties can implement projects in their or other Annex I nations that remove GHGs or create additional carbon sinks. A JI project is then quantified in an emission reduction unit (ERU).[12] JI projects are undertaken by parties in Annex I countries. A JI ERU transfers a unit of allowed carbon emissions from a selling country's cap to the purchasing country's cap. Unlike a CDM CER, which creates an additional emission unit added to the cap, a JI project transfers a credit under the existing cap from one nation to another nation, as a zero-sum transaction. Whereas the CDM process creates additional room in the envelope of permissible carbon emissions by developed nations, the joint implementation process transfers a static quantity of existing allocated credits under the cap from one Annex I developed nation to another. However, JI projects have less burdensome transaction costs than CDM projects, as the former are approved and administered by the parties involved, rather than the UN Kyoto Executive Board, and are not subject to detailed periodic monitoring.

The use of offsets for compliance increases the compliance options, and by increasing supply can decrease total costs of compliance by an estimated 71%.[13] CDM projects may only be pursued in developing nations for use by regulated entities in developed Annex I countries. As of the end of 2006, the World Bank reports that CDM projects were located 61% in China, 12% in India, 7% in other Asian countries, 10% in Latin America (most significantly in Brazil), and 3% in Africa.[14] Africa was largely left out of CDM projects.

CDM CERs and JI ERUs are required to be additional to baseline project emissions.[15] This involves the establishment of an individual emissions baseline, taking into account sector reform initiatives, barriers to expansion, and sector expansion plans.[16] Environmental groups have questioned the additionality of renewable energy projects, if their construction is not because of the value of the offset sale (see chapter 15).

The Kyoto Protocol collects 38 developed nations into a voluntary agreement to limit their carbon emissions. Each of these nations decides how to impose these limitations on its local industries. Those covered emitters of carbon needing additional allowances can either create or purchase additional allowances through the JI or CDM mechanisms.

Thus, the emission cap of any country includes Kyoto credit assigned amount units (AAUs) plus carbon removal units (RMUs) from forestation projects that remove CO_2 from the atmosphere, plus JI ERUs and CDM CERs earned and certified on unregulated carbon reduction projects in developing nations.

Risks and Benefits of Carbon Trades

Emission credit creation and trading can serve as a means to transfer funds for carbon reduction to developing countries, as well as fostering least-cost emission mitigation activities internationally. As mentioned in chapter 3, the CDM apparatus emerged as a last-minute compromise creation of the 1997 Kyoto Conference.[17] It is patterned on the U.S. SO_2 trading experience, now writ multinational. The CDM offset-creation element of the Kyoto Protocol works as a de facto indirect cap on the cost of traded carbon allowances. CDM projects may only be pursued by registration of the credit through the Kyoto Protocol by entities in Annex I countries. Since Kyoto CDM CER offsets must be created in developing countries, there is sovereign risk and commercial risk associated with the creation and registration of these intangible items.[18] These risks are mitigated to some degree by the fact that the international apparatus is overseen by UN designated authorities in each such developing nation that hosts a CDM project.

There also is price risk in trading in these credits. In fact, this commercial risk is present in the trading or speculation in any emission, renewable energy, or carbon allowances or offsets. The value of carbon aggregators of CER credits has plunged, with the share prices of five public carbon market makers and CDM development companies plunging 13–98% from mid-2007 to mid-2008. There also is a timing factor in the creation and trading of approved credits. It takes about 300 days for a CDM project to move in the Kyoto mechanism from the stage of validation to registration. If carbon credits become the biggest market in the world, as expected once all major carbon emitters are involved, the quality of the credits traded becomes a crucial factor. The value of carbon credits and offsets is forecast to increase in value from €8.5 billion currently to €200 billion.[19]

In Kyoto Annex I countries, private markets have arisen with platforms for the trading of offsets. Selling carbon emission credits is typically done through forward contracts. Carbon trading is becoming sophisticated. Structured finance and derivatives are emerging in the carbon trading market as it matures and becomes more pervasive. The CDM CERs and other green credits share some characteristics with other commodity-based asset classes. They are not self-liquidating assets; forward delivery contracts are delivered for the CERs, and the receivable is dependent on the performance of the project generating the CERs. Unlike commodity-based asset classes, there may not be any commodity created backstopping the CO_2e reductions accomplished. The one exception to this is production of renewable energy, which does create the energy commodity or service in addition to the CER credit.[20]

Whereas the CDM process creates additional room in the envelope of permissible carbon emissions by developed nations, the JI process transfers a static quantity of existing allocated credits under the cap from one developed nation to another. Therefore, there is an advantage to the math of CDM investments. Under the Kyoto Protocol CDM, CERs, and JI ERUs can be used in future compliance to satisfy up to 2.5% of the party's annual allowed emissions.[21] CERs and ERUs obtained prior to 2008 can be fully banked for use in the 2008–2012 compliance period. The Kyoto Protocol does not place limits on the use of excess allowances other than that tradable allowances must be supplemental to significant domestic measures to reduce GHG emissions.[22]

Although seen as the world model for cap-and-trade carbon control, the Kyoto Protocol has actually been on the ground for less time than the European Union's own carbon regulatory scheme. The European Union scheme is particularly important for the following reasons:

- The European Union regulation commenced three years before Kyoto commenced
- Of the 38 Kyoto-regulated developed countries, 33 are in Europe

The next chapter dives into the EU regulatory experience with carbon. It showcases not only a maturing scheme of regulation, but a number of schisms and dissents that foreshadow regulatory and governance issues in other countries that will follow suit.

Notes

1. UNFCCC Articles 3(1), 4(1) ("All Parties, taking into account their common but differentiated responsibilities…"). The concept was originally part of the Montreal Protocol on Substances that Deplete the Ozone Layer, Sept. 16, 1987, 1522 U.N.T.S. at Article 5 and at Annex II, paragraph T, Article 10 (as amended in the London Amendments, June 29, 1990, UNEP/OzL.Pro2/3).

2. UNFCCC. 2001. Decision 19/CP.7. Annex of the Marrakech Accords.

3. See, Conference of the Parties Serving as the Meeting of the Parties to the Kyoto Protocol. 2006. *Modalities and Procedures for Afforestation and Reforestation Project Activities Under the Clean Development Mechanism in the First Commitment Period of the Kyoto Protocol*. Decision 5/CMP.1 March 30. http://cdm.unfccc.int/Reference/COPMOP/08a01.pdf#page=61.

4. Kyoto Protocol to the United Nations Framework Convention on Climate Change. 1998. Art. 12, sec. 37, I.L.M. 22.

5. Kyoto Protocol, art. 12, sec. 5.

6. Report of the World Commission on Environment and Development. United Nations General Assembly Resolution 42/187, 11 December 1987.

7. See, UNFCCC. 2006. Decision 5/CMP.1, UN Doc FCCC/KP/CMP/2005/8/Add.1.

8. Hart, Craig. 2007. The Clean Development Mechanism: Considerations for Investors and Policymakers. *Sustainable Development Law & Policy*. American University, Washington College of Law. 54 (Spring): 46. (utilizing UNEP data).

9. Elgin, Ben. 2007. Little Green Lies. *Business Week*. October 29: 45.

10. Wara, Michael. 2008. Measuring the Clean Development Mechanism's Performance and Potential. *UCLA Law Review*. 55 (August): 1,785.

11. El-Ashry, M.T. Framework for a Post0Kyoto Climate Change Agreement, American University, Washington College of Law Sustainable Development Law & Policy, Winter 2008.

12. Kyoto Protocol, art. 6, sec. 37. I.L.M. 22.

13. U.S. Government Accountability Office. 2008. *Carbon Offsets: The U.S. Voluntary Market Is Growing, but Quality Assurances Poses Challenges for Market Participants*. GAO-08-1048. (August): 33.

14. Etter, Lauren. 2007. In China, a Plan to Turn Rice into Carbon Credits. *Wall Street Journal*. October 9: A1, A15.

15. Kyoto Protocol, art. 3, 5, and 7.

16. CDM Modalities and Procedures, art. 48.

17. Kyoto Protocol, art. 12.5 and 6.

18 For treatment of sovereign risk and commercial in developing countries, see Ferrey, Steven. 2010. *The Law of Independent Power*. Eagan, Minnesota: Thomson/West Publishing. Section 3:10. Commercial risk is mitigated by pooling CERs from different technologies and from different countries in financial instruments for trading.

19 Carbon Rating Agency. 2008. *Carbon Ratings*. June 5.

20 For a discussion of whether electricity production is a good or a services, see Ferrey, Steven. 2004. Inverting Choice of Law in the Wired Universe: Thermodynamics, Mass and Energy. *William & Mary Law Review*. 45: 1839.

21 Kyoto Protocol, art. 12, sec. 37. I.L.M. 22.

22 UN Conference of the Parties. 2001. Framework Convention on Climate Change. Seventh Session, Marrakech, Morocco. October 29.

6. THE EUROPEAN UNION CORE OF CARBON CONTROL: COMPARED AND CONTRASTED WITH RECENT U.S. EXPERIENCE

The European Union Carbon Basics

Europe plays a critical role in carbon regulation. The European Union (EU) monitors and regulates CO_2 as its EU compliance component consistent with the larger Kyoto Protocol. As a condition of EU membership all EU members must comply. The EU carbon scheme monitors only about half of CO_2 emissions—those that are more easily ascertainable and tracked from certain large industries and downstream sources. To account for the other half of CO_2 emissions, regulatory systems would have to go upstream in the production process and try to monitor or assess CO_2 emissions from smaller sources. The EU system of regulation includes any combustion power source exceeding 20 MW, designed to exclude households and the agricultural sector.[1] It also excludes several sectors that are major emitters of CO_2, such as transportation and aviation.

EU members from 27 countries constitute the core group of the 38 Annex I Kyoto Protocol countries whose carbon emissions are regulated. In addition to the European countries that are members of the EU, there are several non-EU European countries that are covered in the Kyoto Protocol. The EU Emission Trading Scheme (EU-ETS) covers approximately 5,000 European-based companies and 12,000 sources of industrial site emissions.

The EU regulations required that, by July 1, 2004, all business consumers in EU countries should have been able to get their electricity from their choice of suppliers, and that beginning July 1, 2007, all residential consumers will be able to choose among deregulated suppliers. As of 2005, 18 of the 25 members of the European Union still had not enacted legislation on electricity deregulation.

The current EU carbon regime reflects as much political trading as an objective application of neutral scientific principles. All EU countries are not treated similarly. The 27 participating Kyoto Protocol Annex I EU countries made significant political differentiation among their responsibilities to reduce carbon emissions ranging from a 28% carbon reduction (Luxembourg) to an allowed 27% carbon increase (Portugal). Central and Eastern European states have launched legal proceedings against the European Commission, alleging their assigned amount unit (AAU) allocations are now too low. Outside the European Union among the Kyoto Protocol Annex I countries, Australia is allowed to increase emissions up to 8%, while Russia, Ukraine, and New Zealand have no reduction requirements.

Phase II of the European Union Emission Trading Scheme (EU-ETS) corresponds to the Kyoto Protocol 2008–2012 initial phase. EU-ETS allowances (EUAs) are given away by the 27 EU-ETS countries to their industries as part of their internal political processes. The Kyoto Protocol has distributed AAU allowances without charge within both the Kyoto and the European Union carbon schemes.

Trading

Trading of emission rights is allowed[2] under the EU-ETS and the Kyoto Protocol.[3] Unlike the more comprehensive Kyoto Protocol, the EU-ETS only covers CO_2 among GHGs. For the EU trading market, trading in EU-ETS CO_2 allowances hit $30 billion in 2006, according to the World Bank.[4] Any party can purchase EUAs, even if it does not itself require them for compliance. This includes those traders who wish to speculate in these regulatory commodities.

During Phase I of the EU-ETS prior to 2008, the penalty for an allowance shortfall by a regulated CO_2 emission source was €40/CO_2e plus market purchase of emissions allowances to cover the deficit. During the current Phase II of Kyoto covering 2008–2012, the penalty has risen to €100/CO_2e plus purchase of enough additional to cover the deficit.[5] This is a 150% rise in the penalty assessment for the current phase of the EU carbon regulatory system. By contrast, there is no penalty for noncompliance with Kyoto Protocol requirements.

Trading prices of carbon allowances and credits can be volatile. When it was announced that verified emissions were 41 million metric tons of CO_2 equivalent, or approximately 2.5% lower than expected, the price of the EUA credit market of the EU-ETS system plunged 67%.[6] The price of EUAs fell from €30 to virtually nothing so quickly, before rebounding, because of allocation imperfections. With the European target for reductions of carbon squarely on the utility sector, it was allocated a deficient number of credits, creating a shortage; while other industries, such as cement, refineries, etc. were over-allocated credits by their host countries. These latter industries were slow to sell their allocated EUAs. But when they did, it created a surplus that flooded the market and drove down trading prices.

The EU-ETS system allows banking of allowances (although not from Phase I ending in 2007 carried over to Phase II beginning in 2008). This system works to dampen any short-term price swings in allowance or offset availability and trade pricing. The EU program also allows borrowing of next-year allowances to satisfy current-year allowance requirements. Because the actual CO_2 reductions were relatively modest, there is no significant bank of excess CO_2 credits to buffer this market. This is in contrast to the U.S. SO_2 allowance program that created a large bank of surplus annual emission credits that buffered the market and allowed it to enjoy a relatively constant price. The European Union is proposing to further limit emission allowances granted to renewable energy projects.

Linkage with Kyoto

The EU Linking Directive links the EU-ETS system with the Kyoto Protocol JI and CDM mechanisms.[7] Both CERs (created as offsets under the Kyoto Protocol CDM) and ERUs (created under the Kyoto Protocol JI program) are fully fungible in Europe. The EUAs can be traded and banked until 2012 within the EU-ETS for purposes of EU carbon compliance. Kyoto CDM offset credits have become the offset mechanism for EU-ETS additional credits. So far there is little linkage between the major trading programs in non-EU countries. The EU-ETS is linked with other European programs in the countries of Iceland, Lichtenstein, and Norway, which are not EU members, but participate in the Kyoto Protocol. Contracts for differences can be employed to hedge future carbon compliance requirements in the European Union.

While the EU-ETS is linked to the Kyoto Protocol, there are differences.[8] CERs from Kyoto land-use and forestry projects cannot be applied as EU-ETS credits. And there are percentage limitations on compliance use of CERs.[9] It is still unresolved whether any federal or state carbon regulations in the United States will link with the EU or Kyoto schemes, although California is exploring these options (see chapter 8). If a *trigger* threshold in allowance price is exceeded within the East Coast U.S. Regional Greenhouse Gas Initiative (RGGI) scheme, Kyoto and EU offsets become eligible in this U.S. state carbon regulation.

For an allowance or offset trade, the European Union is moving toward recognition of the taxable location of such trade being the location of the buyer. While there is now no cross-border taxation on allowance trades, there will be a substantial trade into the European Union of CERs from outside the region. This movement toward buyer location localizes a tax revenue-raising locus inside the Annex I countries of Europe. The tax implications and loci of the quickly increasing volume of carbon trades becomes a critical competitive issue. This market is expected to become one of the largest and most vibrant trading markets in the world in short order. Tax implications and tax minimization strategies become more central.

The lesser price at which CERs have traded compared to actual EU-ETS EUA trade prices creates arbitrage opportunities for stakeholders in the market. There has been a market created in the difference in prices between EU-ETS EUAs and CDM CERs, which are freely utilized within the EU-ETS as EU offsets, even though created under the Kyoto Protocol rather than under the EU-ETS. CERs are fungible to add to one's cache of creditable emission allowances. Because of a more established market in the larger EU-ETS market, the restrictions in some nations regarding the number of CERs that may be utilized by regulated entities annually have tended to trade in Europe at about a 20–25% discount to the price of allowance units. This is influenced from the uncertainty as to whether CERs will continue after the now-scheduled 2012 termination of the Kyoto Protocol. However, for entities complying now during the second phase of Kyoto with a combination of EU-ETS EUA units and CERs, they can purchase CERs and sell EUA units, realizing a 20–25% gain on the swap, and not changing their net held allowance credits.

The Future of Carbon

The post-2012 EU Emissions Trading Scheme is starting to be scoped in EU member meetings. The EU-ETS utilizes National Allocation Plans for the (free) distribution of carbon emission allowances.[10] The quantity of allowances a nation can issue is governed by 12 EU-ETS criteria, but otherwise national discretion is not explicitly prescribed by the European Union. Because of inconsistencies and controversies in individual countries, there is general agreement on post-2012 centralized EU allocation of emissions rights, eliminating national allocation. In its second allocation round for EUA national allowances to emit CO_2, the European Commission plans to centrally determine future national allowances. Post-2012, the European Commission's proposal is for auctions of allowances to be phased in, rising to 100% auction by 2020, with possible 100% auction for power generators starting in 2013. The estimate is that such auction will generate €30–50 billion ($45–75 billion) annually. In a time of economic uncertainty, additional revenue sources are attractive.

The auction of allowances in the EU system has never been generally used before for any environmental attribute. In fact, it has never been used before in the U.S. system, until the start of the RGGI carbon regulation in 2009 (see chapter 7). In the United States, auction of carbon allowances is raising constitutional and other legal issues (see chapters 13–14). However, once outside the constitutional separation of regulation in the U.S. system, there is freedom in other countries for auction of carbon allowances.

CDM CERs created pre-2013 would be allowed to be carried over until 2020, under current European Commission proposals. The European Commission also is debating whether offsets, currently administered through the CDM system, will continue to be recognized after 2012. Currently debated concepts would allow up to one-third of EU-ETS emissions reductions to be satisfied by foreign-based CDM project CERs through 2020, rather than through reductions of carbon emissions occurring in the European Union. "Europe is reducing more emissions outside than inside," which could "displace national policies" in countries like China, which harbor more than one-third of all CDM projects and may lack incentives to establish additional carbon reduction policies, according to Juan Delgado,[11] climate specialist in Brussels.

A number of European countries are administering their carbon systems differently. For example, the European Union ETS for registering compliance, designed to be synched with the Kyoto Protocol, does not use actual measurements but rather calculates emissions and carbon sequestration based on proxy values. There is a tendency for estimation error in the use of proxies. For the first 175 CDM projects that issued CERs, the validation procedure overestimated the number of CERs produced by approximately 27% on average, with a standard deviation error of 42.5%.[12] This was a mechanism for both host CDM countries and CDM investors to inflate the number of CERs created, traded, and sold.[13]

The fact that compliance was shown so easily in 2006 in the EU-ETS, after early predictions of shortfall, indicates that there is measurement flexibility.[14] The change of GHG emissions by regions is shown in figure 6–1. With lack of enforceability, those major industrialized nations are not even meeting the relatively modest targets for the first commitment period of the Kyoto Protocol. By contrast, the U.S. system of emissions regulation and trading of SO_2 credits requires actual verification of emission reductions.

The U.S. Congress Government Accountability Office (GAO) in November 2008 reported to the Congress on the lesson of the EU-ETS:[15]

- The high cost of producing CDM offsets and less costly options

- The failure of up to 40% of CDM offsets to meet *additionality* requirements in fact (see chapter 15)

- Little positive impact on sustainable development (see chapter 10)

- The failure of undesired leakage of carbon, protected by free allocation of allowances rather than auction of allowances (see chapter 13)

The EU-ETS carbon regulation was a parallel early start for the now 27 EU-member countries and three other participating European non-EU countries (Norway, Iceland, and Liechtenstein) that also are covered by the Kyoto Protocol. The European Union is the only region of the world prior to 2008 to have implemented a cap-and-trade system for GHGs.[16] However, twenty-three U.S. states joined by four Canadian provinces, under four different regulatory regimes, also are moving forward on cap-and-trade carbon regulation (see chapters 7–9).

Total Greenhouse Gas Emissions by Region

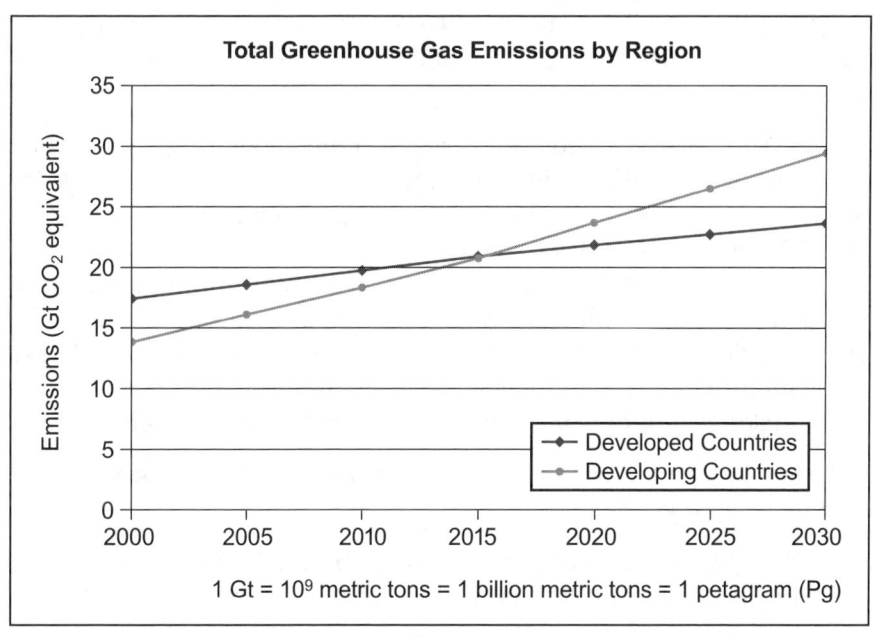

Fig. 6–1. Regional GHG emissions between developed and undeveloped countries

Tensions within the European Union Compared to Plans in the United States

Despite Kyoto and its binding requirements on EU countries, European GHG emissions in industrialized European countries are increasing.[17] Many EU countries are forecast to miss their Kyoto targets, with the exception of two former Soviet countries. Only Russia and Poland among the developed countries covered by the Kyoto Protocol are expected to satisfy their 2010 targets, and this is because of the post-Soviet Union economic collapse in those countries which has shuttered many existing CO_2 emission sources.

Most excess Kyoto allocated emission allowances are held by Russia and Ukraine. These are expected to be in excess of 100 million metric tons of CO_2 equivalent per year.[18] These excess emission allowances are approximately 33% of the validated Clean Development Mechanism emissions reductions as of May 1, 2007. They account for almost one-half the number of CERs expected to be issued assuming a validation estimate error of 27%. Aside from EU countries, Canada and Japan are projected to miss their interim 2010 CO_2 targets by 500 million tons each; Japan's emissions are rising, and Canada has backed away from its target obligations.[19]

The World Wildlife Federation accused the EU of "playing tricks on the atmosphere" with the EU-ETS. The Federation claimed that a 4–5% reduction from 2009 onward will achieve the planned EU 20% reduction by 2020, because 8% is accounted for by deindustrialization in the former Soviet states, and an additional 8% will be achieved through the CDM mechanism in foreign developing nations. According to this criticism, by achieving domestically only one-quarter of the 20% goal, the EU will take credit for the entire 20% reduction. The Federation also claims that the EU is not making additional in-country carbon reductions (see chapter 15).

There are a handful of recent major serious policy disputes within the European Union:

- Whether allowances to emit CO_2 will be auctioned to industry beginning in 2013, or freely allocated to industry
- Conflict between original and later-joining EU member states
- The means of future restrictions application to coal-fired electric power production
- Whether states are willing to cede decision making to a central EU authority
- The speed and base of program requirements

Auction of allowances

Auction of allowances emerged as a key conflict. If auction of allowances is adopted, it has profound wealth-shifting implications. The EU-ETS program originally committed to free allocation of allowances to regulated industrial emitters. Even though in Phase I (2005–2007) up to 5% of carbon emissions allowances were allowed to be auctioned by an EU country, only four of the 27 nations employed any auction (Denmark, Ireland, Hungary, and Lithuania), and they in total auctioned only 0.13% of total allowances.[20] During the current 2008–2012 Phase II, it is expected that eight Western European nations will auction about 3% of EU-ETS allowances.[21] Auction of CO_2 allowances to all power generators has been proposed by the European Commission to commence in 2013, phased to 100% auction by 2020. There is a push post-2012 for centralized EU allocation of emissions rights, eliminating national allocation.

Poland and other Eastern EU nations militate against immediate auction of allowances. They assert that their industries are more energy-intensive than in Western EU nations, and thus would be competitively disadvantaged by allowance auction; France and Luxembourg join this concern. Italian leaders resisted EU auction of allowances as "unsuitable . . . untenable" and "an act of madness."[22] Political concessions were made in Poznan, Poland, meetings in December 2008, where the former Eastern-block EU states were given 12% of future auction revenues from allowance auction, for them to subsidize continuation of certain high-carbon industries.

The 2008 financial meltdown has increased pressure to raise revenues through auction rather than traditional allocation of allowances without charge. That pressure is felt both in the European Union and the United States. Environmental groups have sought auction of all allowances to raise revenues in both the European Union and the United States.

In the United States, there also has been a divergence on auctions. In the first U.S. carbon program, RGGI (which commenced in January 2009), the guiding principles agreement provided, "The initial phase of the cap-and-trade program will entail the allocation and trading of carbon dioxide allowances to and by sources in the power sector only."[23] However, the RGGI states have since switched to an auction of allowances to essentially any bidder, not an allocation to affected facilities as contemplated in the guiding principles agreement (see chapter 7). New York also agreed to cede some of its allowances to other states to achieve consensus to participate in RGGI. This will transfer income among states.

Issues in the United States on carbon regulation are mirroring some of those in the European Union. Sempra Energy Utilities, the electric and gas distribution utility in the greater San Diego area, questioned the legality of auctioning allowances if the revenues from there were returned to the state general fund. The Los Angeles Department of Water and Power (LADWP), the largest municipal utility in the United States, threatened legal suit over the California proposal to auction carbon emission allowances, alleging that it would result in a $1 billion/year transfer from legacy coal utilities in the southern part of the state to northern legacy non-coal utilities and their ratepayer in the north. They also charged that auction of allowances was an illegal tax and violated the state constitution. The coal, power, and railroad industries threatened some states with suit over the RGGI auction program in the eastern United States. Auctioning state carbon allowances becomes even more contentious in the United States, as it raises issues of possible unconstitutionality of states crossing the *bright line* that separates federal and state energy regulatory authority, depending on implementation, within the U.S. legal system (see chapters 13–14).

Conflict with new entrants and degree of development

The carbon tensions between nations in the European Union mirror each nation's degree of development and economic self-interest, notwithstanding that impact of each molecule of CO_2 has identical global impact. Europe is the first regional carbon trading area of the world, and contains 33 of the 38 Kyoto-regulated Annex I carbon-emitting nations. The five other Kyoto carbon-regulated countries have no common borders with other countries (Japan, Australia, and New Zealand) or border only with non-participating countries (Canada and Turkey).

The 27 EU member nations are increasingly more acrimoniously divided on carbon policy. At odds are the European Union's 15 more affluent original Western European member states, and the more recent, less affluent former Soviet block Eastern European members—eight of whom joined the EU in 2004 and two more in 2007. The EU-ETS carbon scheme originally commenced its Phase I in 2005, covering 25 EU countries. In 2007, the European Union added Bulgaria and Romania and raised the number to 27, with all required to participate in CO_2 reduction through the EU-ETS. In addition, three non-EU member countries (Norway, Iceland, and Liechtenstein) recently joined the EU-ETS carbon system, but are not members of the European Union.

Romania and Bulgaria demanded larger CO_2 emission allocations than they were given by the European Commission, and thereafter judicially appealed this allocation to the European Court of First Instance.[24] During Phase I of the EU-ETS (2005–2007), these Eastern block countries received an allocation surplus of free allowances based on a 1990 baseline pre-Soviet restructuring, which allowed their industries to sell surplus free allowances for profits. In the current EU-ETS Phase II (2008–2012), these countries have had their CO_2 allocations slashed. However, post 2012, the EU proposal, reflecting accession to political pressure to achieve consensus, now is to allow central and eastern EU countries to increase emissions up to 20% above 2005 levels, rather than reduce carbon.

The eight Eastern European EU countries, joined by Italy and Greece, are the 10 countries leading the revolt on policy inside the EU-ETS. Poland, the Czech Republic, Hungary, Romania, and Slovakia launched legal actions against the EU Commission, asserting that even their relatively modest future carbon limits are too strict for the countries' economic growth.[25] These five countries, along with Estonia, Latvia, and Lithuania, form a block of eight countries that want more concessions on future carbon regulation.

These eight countries, joined by Greece, representing one-third of EU members, have a combined voting power to block or stall EU carbon control action. The EU process allows at least 91 votes of the 345 in the European Council to block further actions, just as 40% of U.S. Senators can block action on carbon legislation. Prime Minister Ivars Godmanis of Latvia stated that he would veto any climate package unless it contained more concessions for the Eastern countries that joined the EU in 2004.[26] "Poland is ready to veto" stated Poland Foreign Minister Radek Sikorksi.[27] Czech President Vaclav Klaus has publicly questioned whether there is such a thing as human-made climate change. Less developed areas have resisted consensus.

There is some precedent for phasing in the compliance obligation of developing countries with international environmental requirements. The Montreal Protocol did include trade sanctions for enforcement, giving developing countries a 10-year grace period. Some countries are now discussing trade sanctions as a mechanism to force reluctant countries to adopt requirements. However, tariff sanctions could run afoul of World Trade Organization (WTO) requirements against trade barriers concerning renewable energy sources. Article XX of the General Agreement on Tariffs and Trades (GATT) provides health and environmental exceptions that can even be pursued outside national territory if aimed at a conservation or protection goal. Even the Kyoto Protocol states that its mechanisms should not imperil international trade.[28]

For comparison to issues in the United States, there is a similar theme emerging. The LADWP continued to fight the California cap-and-trade proposal, arguing it as a wealth transfer between utility ratepayers in different parts of the state, and voicing distrusts that auction funds once in the hands of state legislators would be returned to utility ratepayers. Those states that will be last to embrace carbon regulation in the United States have a tremendous ability, if they resist carbon controls, to be the *leakage* in the system (see chapter 13). The results of modeling commissioned by the RGGI Staff Working Group found that a substantial proportion of CO_2 emissions avoided by RGGI will be offset by corresponding increases in non-RGGI states, such as Pennsylvania (see chapter 7).

Embedded high-carbon coal-powered generation

This Eastern European EU block joined by Greece and Italy fear power sector uncompetitiveness or leakage of other power into their economies vis-à-vis larger utilities in France and Germany. To reduce annual carbon emissions, if EU countries are forced to shift from coal to less carbon-intensive natural gas-fired power generation, it would make EU countries more dependent on natural gas imported from Russia. According to Slovakia foreign ministry spokesman Jan Skoda, "We are dependent on Russia for 97% of our gas and more than 90% of our petrol."[29] Fresh in memory is Russia's quick strike into neighboring Georgia in August 2008, and its termination of gas supplies to the Ukraine in January 2006 and again in 2009, which greatly diminished gas supplies throughout the European Union.

"Fifty-eight percent of the world's gas is owned by Russia, Iran, and Qatar. Coal is on every continent," notes coal executive Richard Budge.[30] Poland generates 95% of its power production from coal, which will require more auctioned CO_2 allowances than other fossil fuels, and could increase Poland's electric power prices by up to an estimated 90%.[31] This resistance resulted in recent EU fractious concessions to the most coal-burning nations, to hold the EU-ETS together. At the Poznan, Poland, Kyoto Protocol carbon meetings in December 2008, EU nations tentatively agreed to extend the allocation of free carbon emission allowances to utilities equal to the needs of an efficient coal-burning power plant and doubled the time by extending compliance dates to 2020. This threatens even the modest goals of the Kyoto Protocol and the EU-ETS, and contrasts with the recent call by some members of the scientific community for radical reductions in CO_2 emissions before a 2015 tipping point in global climate (see chapter 4).

Similar divisions to those in the European Union are discernible in some of the states in the United States considering carbon regulation. The areas of the United States most reliant on coal power are not the leaders in U.S. carbon regulation. Disputes have already erupted between California and the other six Western Climate Initiative (WCI) member states, still years before program commencement, over whether the nature of the cap is too restrictive to the California power industry, which is the first industry sector targeted. The Midwest states carbon program has stalled amid dissent involving coal-dependent states (see chapter 9).

Similarly, the U.S. states that have lower relative median incomes also are not generally the leaders on carbon regulation. The president of

Duke Energy worries that federal carbon regulation will be "California-centric" and "ideologically driven."[32] He claimed that California imports lots of coal-fired and hydroelectric power, has the highest U.S. disposable income, and that California has high levels of electric consumption.

The coal industry has threatened suit against RGGI as an unauthorized tax or otherwise illegal.[33] A New York RGGI official commented that there is a substantial chance of litigation challenge in New York, Maryland, and Massachusetts.[34] This prophecy was realized in 2009, when the RGGI carbon regulatory program was successfully challenged in New York by an owner of independent power generation that was subject to that regulation. Of note, this was a gas-fired project, rather than coal-fired generation, which would be required to obtain more carbon allowances or offsets, and was a cogeneration qualifying facility pursuant to the U.S. Public Utility Regulatory Policies Act (PURPA). This generation asset owner challenged the RGGI scheme as an improperly authorized state program in violation of the state's constitution, as well as a violation of the Compact Clause of the U.S. Constitution.[35]

Centralized versus decentralized carbon control

There is movement to centralize Brussels EU allocation of carbon reduction targets rather than continue state control. These state-set targets have not resulted in a reduction of internally generated carbon. The current EU carbon regime represents as much political expediency as an objective application of neutral scientific principles. The 27 participating Kyoto Annex I EU countries made significant differentiation among their responsibilities to reduce carbon emissions.

Consequently, five of the more developed EU-ETS countries (United Kingdom, Italy, Spain, Ireland, and Austria) were short of freely-allocated allowances, while several Eastern European EU members were over-allocated allowances. This led to these Eastern EU subsequent allowance trades earning approximately €700 million from electric power industries in Western EU countries.[36] The power industry was the sector that shouldered that shortage, while other industries in the country were protected in national allocations. This was because the often monopolized power sector did not face international competition in most EU countries, and thus did not face leakage of market share from supply outside the state, even though retail deregulation should have occurred.

Because of inconsistencies and controversies in individual countries, the plan for post-2012 is centralized EU allocation of carbon emissions rights, eliminating current national allocation. Yet, some of the Eastern

European countries (Poland, Czechoslovakia, and Hungary) are expected to challenge their future allocations. Central and Eastern EU states have launched legal proceedings against the European Commission, alleging their allocations are already too low.

These issues parallel similar concerns in the United States. To get consensus for formation of RGGI, New York gave away some of its carbon allocation to other states. In RGGI, a deregulated power sector shoulders the entire carbon reduction burden. There are disputes over whether the 23 states starting to regulate carbon will cede authority willingly to federal control, or whether they will be excepted from selective requirements. There is vigorous dispute in the U.S. Congress over whether to grandfather existing state carbon regulation, or centralize a single national system that preempts preexisting state programs either temporarily or permanently. The Markey-Waxman climate control legislation pending in 2009 would preempt all preexisting state programs between 2012 and 2017, although there is no particular justification for this preemption period.

A similar conflict already is visible at the regional level in the United States. The state of California has complained that the Western Climate Initiative will impose an inordinate burden on the California power sector by excluding the transportation sector until 2015 (see chapter 9). Because California utilities rely on out-of-state electricity imports, California utilities argue that they require extra allocation of any allowances from other states. And such conflicts already are dividing stakeholders within states. Terry Tamminen, an energy advisor to California Governor Arnold Schwarzenegger, characterized the LADWP position against California's planned 2012 carbon control as "morally bankrupt...it is time for those utilities [that] have put themselves in this position to step up and internalize the cost that they have been foisting on the rest of us for decades....so that people in Los Angeles can have cheap electricity."[37] Tamminen stated that potential legal challenges could pose the biggest stumbling block to California's climate change initiatives.

The pace and scope of carbon control

The EU decision to move forward in Poznan, Poland, in December 2008 was more of political and economic compromise than principle. A three-year delay was given to the auto industry, and other select industries in key countries. The former Soviet states were given 12% of auction revenues from allowance auction to allow them to subsidize certain of their high-carbon power plants. The European Court of Justice rejected a legal challenge by a steel company of the failure of France to regulate non-steel emitters of CO_2 when it regulated the steel industry.

The rationale of the court seemed expedient, noting that there is a difference in the aggregate emissions of steel and other metals, and that France could treat with disadvantage certain industries.

At issue is whether the original EU and Kyoto Protocol 1990 carbon emission baselines (prior to Soviet collapse) or recently proposed 2005 carbon baselines (reflecting lower Eastern EU CO_2 levels) will be the baseline against which future carbon compliance will be measured. Poland and Bulgaria argue that more advanced Western EU countries should do more carbon reduction, while the poorer Eastern European countries should do less. Poland's prime minister, Donald Tusk, noted that no global warming-related actions should result in an increase in the price of energy, especially during times of economic downturn.[38] Italian Prime Minister Silvio Berlusconi called for less expensive carbon regulation in tough economic times: "I am ready to use our veto powers. Our companies are in no state to take on costs like those we thought about last year."[39] He stated that the proposed EU-ETS carbon reduction targets for 2020 would "crucify Italy."[40]

The debate in the United States follows similar lines. First, national carbon legislation was held up in the U.S. Congress in 2008 regarding technological and policy compromises, and 40% of U.S. Senate members can do so indefinitely if consensus is not achieved. During a period of economic downturn, similar questions are emerging amid the debate as to how aggressive U.S. carbon control can be. California Governor Schwarzenegger, one of the earliest and most stalwart proponents for carbon regulation, moved in early 2009 to ease green power regulations because of California's economic downturn.

The next three chapters survey the U.S. system of carbon control embodied in state and regional programs:

- The Regional Greenhouse Gas Initiative (chapter 7)
- California (chapter 8)
- Voluntary programs, Western states, and Midwestern states (chapter 9)

Notes

1. European Union Parliament. 2003. Directive 2003/87/EC.

2. European Union Parliament, Directive 2003/87/EC. (Amending Council Directive 96/61/EC.)

3. Kyoto Protocol to the United Nations Framework Convention on Climate Change. 1998. Art. 20; and UNFCC. 2001. Decision 19/CP. 7. Annex of the Marrakech Accords. 2001.

4. Ball, Jeffrey. 2007. Kyoto's Caps on Emissions Hit Snag in Marketplace. *Wall Street Journal*. December 3: 1.

5. European Union Parliament, Directive 2003/87/EC.

6. Timmons, Heather. Data Leaks Shake Up Carbon Trade. *New York Times*. May 26: C-1.

7. European Parliament and Council Directive amending Directive 2003/87/EC of the European Parliament, regarding Kyoto Protocol project mechanisms 2003/0173 (COD).

8. European Union Parliament. 2004. The Linking Directive. EU Directive 2004/101/EC; Kyoto Protocol, art. 17.

9. Kyoto Protocol, art. 11(a)(3).

10. European Commission. 2003. EU-ETS Council Directive, art. 10. O.J. (L 275)(EC). Annex III to Directive 2003/87/EC; Council Directive 96/61/EC; COM(2003) 830 final of 7 January 2004.

11. Gardner, Stephen. 2008. EU Parliament, Council Making Progress on Post-2012 Emissions Trading Scheme, *Environmental Reporter*, July 11, 2008: 1417.

12. Hart, Craig. 2007. The Clean Development Mechanism: Considerations for Investors and Policymakers. *Sustainable Development Law & Policy*. American University, Washington College of Law. (Spring): 42 (utilizing UNEP data). As of May 2007, there were more than 1,800 CDM projects that had estimated their emission reductions through the validation process but not all had verified all their CERs.

13. Voigt, Christine. 2008. Is the Clean Development Mechanism Sustainable? *Sustainable Development Law & Policy*. American University, Washington College of Law. (Winter): 15.

14. Ball, Jeffrey. 2007. Kyoto's Caps on Emissions Hit Snag in Marketplace. *Wall Street Journal*. December 3: 1.

15. U.S. Government Accountability Office. 2008. *International Climate Change Programs: Lessons Learned from the European Union's Emissions Trading Scheme and the Kyoto Protocol's Clean Development Mechanism*. GAO-09-151.

16. Convery, F., D. Ellerman, and C. de Perthuis. 2008. The European Carbon Market in Action: Lessons from the First Trading Period. *Global Change Science Policy Report No. 162*. Massachusetts Institute of Technology. (June): 7.

17 UNFCCC. 2006. *Greenhouse Gas Data, 2006.* (showing an 11-ton increase in GHG emissions from 1990-2004, excluding the former Soviet countries.)

18 Hart, *The Clean Development Mechanism*, 41, 44.

19 Ball, Jeffrey. 2007. Kyoto's Caps on Emissions Hit Snag in Marketplace. *Wall Street Journal.* December 3: A-1, 19.

20 Convery, Ellerman, and Perthuis, *The European Carbon Market in Action*, 11.

21 Ellerman, Denny and P. Joskow. 2008. *The European Union's Emissions Trading System in Perspective.* Pew Center on Global Climate Change. (May): 38.

22 Goldirova, Renata. 2008. Rocky Path ahead for EU Green Legislation. *EUobserver.com.* October 21. http://euobserver.com.

23 See, Regional Greenhouse Gas Initiative. 2003. *Goals, Proposed Tasks, Short-Term Action Items.* 1. http://www.rggi.org/docs/actionplanfinal.pdf.

24 Convery, Ellerman, and Perthuis, *The European Carbon Market in Action*, 22.

25 Herrero-Martinez, Erica. 2007. States Study Carbon Trading. *Wall Street Journal.* August 1: B5A.

26 Castle, Stephen. 2008. European Nations, Fearing Downturn, Seek to Revise Agreement on Emissions Cut. *New York Times.* October 17: A6.

27 Phillips, Leigh. 2008. Italy, Poland Threaten to Veto EU Climate Package. *EUobserver.com.* October 16. http://euobserver.com.

28 Kyoto Protocol, art. 2.3 and 3.5.

29 Levitin, Michael. 2008. Poland Leads Revolt against EU Climate Change Deal. *London Daily Telegraph.* October 4: 19.

30 Landler, Mark. 2006. The Energy Challenge: Sooty Future for Europe, A Green Self-Image Clashes with a Reliance on Coal. *New York Times.* June 20: A1.

31 Hall, Siobahn. 2008. Tough Negotiations on EU Climate Package. *Energy Economist.* Nov. 1, 2008: 33.

32 Ryser, Jeff. 2009. Duke's Rogers Fears CO_2 Legislation Will Be 'California-centric' and Driven by Ideology. *Electric Utility Week.* February 16: 21.

33 Carbon Control News. 2008. First RGGI Allowance Auction May Trigger Coal Industry Lawsuits. *carboncontrolnews.com.* July 21: 1 and 4.

34 Ibid, 5.

35 *Indeck-Corinth L.P. v. David Paterson, et al.*, filed in Supreme Court of State of New York, Sartatoga County (January 2009).

36 Convery, Ellerman, and Perthuis, *The European Carbon Market in Action*, 14.

37 Weinzimer, Lisa. 2008. Schwarzenegger Advisor Says States, Regions Will Take Lead on Climate Program. *Platts Electric Utility Week.* June 16: 7–8.

38 Goldirova, Renata. 2008. Poland Gears Up for Battle over CO_2 Emissions. *EUobserver.com*. October 9. http://euobserver.com.

39 Hall, Siobahn. 2008. Tough Negotiations on EU Climate Package. *Energy Economist*. Nov. 1, 2008: 33; Joe Murphy. Cost Row Threatens EU Climate Strategy. *London Evening Standard*. Oct 16: 2.

40 Charter, David and Rory Watson. 2008. Black Clouds Hang over Green Targets as EU States Say We Can't Afford Them. *Times*. (U.K.) October 17: 6.

7) THE REGIONAL GREENHOUSE GAS INITIATIVE: THE ORIGINAL U.S. REGULATION

To fill the vacuum left by the United States' refusal to participate in the Kyoto Protocol, many states have taken direct regulatory action. Beginning in April 2003, Governor George Pataki of New York initiated the effort by inviting neighboring states to participate in a regional cap-and-trade emissions program. On September 29, 2003, the executive branch environmental agency heads of six northeastern states, including New York, agreed on both guiding principles and an initial timeline for development of the Regional Greenhouse Gas Initiative (RGGI) program.

The guiding principles agreement provided: "The initial phase of the cap and trade program will entail the allocation and trading of carbon dioxide allowances to and by sources in the power sector only."[1] However, all 10 states implementing RGGI offer an auction of allowances to essentially any bidder, not an allocation to affected facilities as contemplated in the guiding principles agreement. This change, as discussed in chapter 14, raises a key constitutional legal issue.

Goals and Regulatory Mechanisms of RGGI

On December 20, 2005, seven states—Connecticut, Delaware, Maine, New Hampshire, New Jersey, New York, and Vermont—entered into an agreement to implement the Regional Greenhouse Gas Initiative.[2] Since that time, Massachusetts, Maryland, and Rhode Island have entered the

RGGI memorandum of understanding (MOU) (collectively, RGGI states). The principal goal of the MOU is for RGGI states to do the following:

> Commit to propose for legislative and/or regulatory approval a CO_2 Budget Trading Program (the Program) aimed at stabilizing and then reducing CO_2 emissions within the Signatory States, and implementing a regional CO_2 emissions budget and allowance trading program that will regulate CO_2 emissions from fossil fuel-fired electricity generating units having a rated capacity equal to or greater than 25 megawatts.

The market-based design of the RGGI MOU is a cap-and-trade program. "Cap-and-trade systems operate by capping the amount of [CO_2] emissions allowances, distributing emissions allowances to sources up to the cap, and requiring each covered source to have sufficient allowances to cover its [CO_2] emissions at the end of each compliance period."[3] This is a supply-side initiative: "CO_2 emission allowances will be allocated to, and traded among, fossil fuel-fired electricity generators within the region that supply electricity to the grid."[4] The first emissions trading programs in the United States included those applied to the phase-down of leaded gasoline in the 1980s, compliance with the Montreal Protocol for reduction of ozone depletion, reduction of SO_2 emissions in the 1990 Clean Air Act, the 1994 Regional Clean Air Incentives Market (RECLAIM) program in Southern California to reduce NO_x and SO_2 emissions, and the EPA NO_x reduction in 12 northeast states under the Ozone Transport Commission program.[5] Recent SO_2 and NO_x allowance trading prices are shown in table 7–1.

The RGGI Staff Working Group (SWG) finalized the Draft Model Rule (Model Rule) in January 2007. The Model Rule is the product of over two years of work by the SWG and it serves as the foundation upon which the RGGI states will base their individual model rules. The Model Rule will be used by each state as a starting point for obtaining regulatory or legislative approval of its cap-and-trade carbon program.

The RGGI MOU set the start date for the program in 2009, making it the earliest program in the United States. In 2009, CO_2 emissions from power plants in the region are capped at historic levels and the cap will remain in place until 2015. RGGI states then begin the process of incrementally reducing power plant emissions, with the goal of achieving a 10% reduction by 2018.

SO₂ and NOₓ Allowance Spot Prices

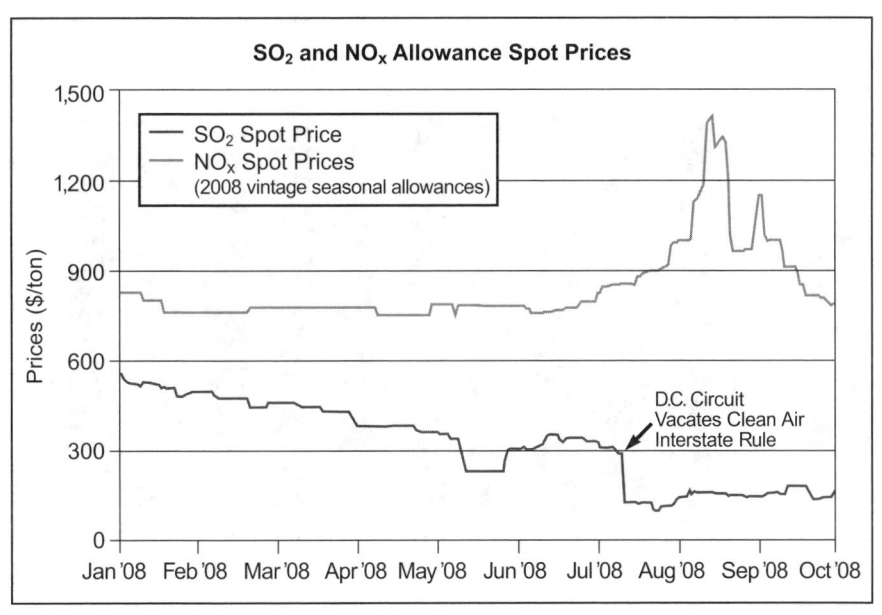

Fig. 7–1. SO_2 and NO_x allowance spot prices. *Source: Derived from Cantor Fitzgerald data*

The MOU creates an annual regional CO_2 emissions budget for the years 2009 through 2014, apportioned among the participating RGGI states. Beginning in 2015, each RGGI state's annual CO_2 emissions budget would decline by 2.5% per year "so that each state's base annual emissions budget for 2018 will be 10% below its initial base annual CO_2 emissions budget." This is shown in figure 7-2. By 2020, the program is expected to reach an emissions reduction of approximately 35% compared to a business-as-usual unregulated scenario.

Each individual allowance provides a limited authorization to emit one ton of CO_2 during the preceding control period. Each control period runs for three years—with the first control period commencing on January 1, 2009 and ending on December 31, 2011. The MOU allows affected facilities to take credit for CO_2 emissions reductions achieved prior to the onset of the RGGI program (i.e., prior to 2009), which will thereby create additional allowances. In addition, allowances, offset allowances, and early reductions may be carried over to subsequent years (i.e., "banked").

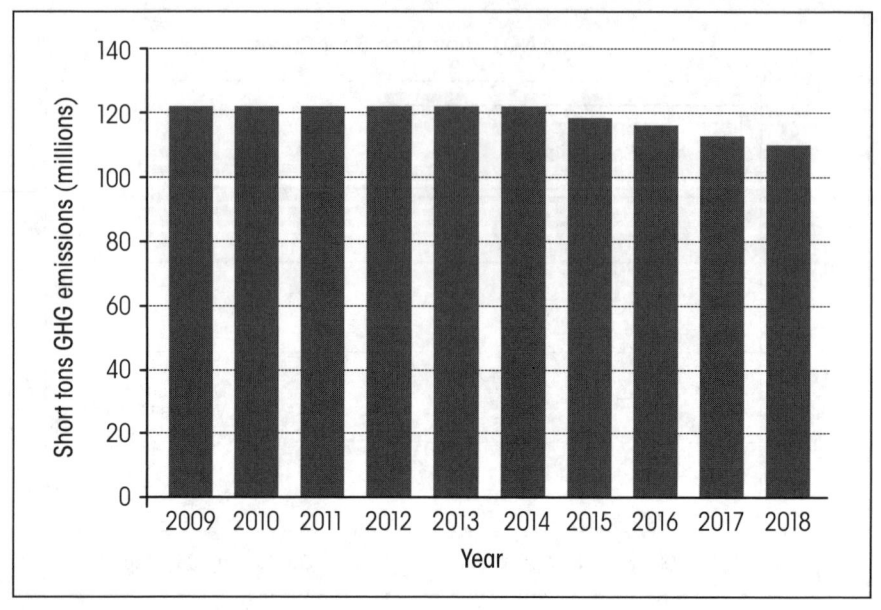

Fig. 7–2. RGGI emissions reductions

Creation and Auction of RGGI Carbon Allowances

One significant aspect of the Model Rule is its requirement that each state reserve a minimum of 25% of that state's carbon allowances for "consumer benefit or strategic energy purpose[s]."[6] Depending on the market for carbon allowances, this could leave states with millions of dollars in an essentially open-ended fund. Consumer benefits could range from supplementing consumer electricity bills or funding state-run energy efficiency programs, to putting the money back into state coffers. In reaction to the "consumer benefit or strategic energy purpose" requirement, power producers lobbied states to only auction the minimum of 25% of carbon allowances and to allocate the remaining shares to power producers based on their historical or future energy production levels without charging for these allocations. It is unprecedented in U.S. environmental regulation history that the allocations for emissions be auctioned rather than given to existing sources. The scale of an auction of all allocations is unprecedented in the world.

There also is little experience internationally for auctioning any emission allowances. For the history of the European Union (EU) carbon program, allowances have been given away for free (see chapter 6), mirroring the U.S. Clean Air Act emission allowance programs in which almost all allowances have been given away to regulated emitters without charge. The EU European Trading Scheme utilizes national allocation plans for the free distribution of allowances.[7] This system covers only EU CO_2 and to date less than 50% of EU emissions. Additionally, no EU country is allowed in Phase II (until 2013) to auction more than 5% of its allowances. After 2012, the European Union is likely to shift to an auction of all power sector allowances in the European Union, eliminating all free traditional allocation by 2020. This will build on the earlier decisions in RGGI in the United States.

Some observers have noted that even five years ago, auctioning allowances for emissions was thought to be a "crazy idea." All of that has changed with the start of RGGI. Some have argued that allowances should be auctioned in the RGGI system because a state government has the right to control the quality of its common airs, and doing so internalizes all of the costs inside the auction system.

Forcing power producers to pay for all of their carbon allowances could, however, create a competitive disadvantage for in-state producers if neighboring states' generators are given allowances without charge or are not required to procure carbon allowances. All ten original RGGI states have independent system operators (ISOs) or regional transmission organizations (RTOs), so there are competitive wholesale markets that will have to incorporate the cost of obtaining carbon allowances. According to a 2008 analysis by the staff of the House of Representatives Commerce Committee, a "tough state program 'may just shift the location of, rather than decrease, national emissions because the sources subject to the more stringent state program will need fewer allowances, thus freeing up allowances for sources in other states.'"[8]

Power producers also expressed their concerns about how this new expense will affect the unchangeable long-term power contracts that they signed prior to the carbon regulation requirement. The cost of future CO_2 allowances was not factored into any of these existing contracts, and generators producing under these long-term deals fear that they will not be able to adjust the established contract prices to account for them. Note that in wholesale markets in restructured states, the contract rights to dispatch a particular power plant typically are transferred to and controlled by an ISO, an entity that will not itself be subject to imposition of carbon costs. The 10 RGGI states and California, which also is leading in carbon regulation (see chapter 8), utilize such ISO entities. Therefore, the party deciding under existing legal contract to operate a generating plant will not be the party that directly internalizes or perceives the marginal carbon cost of operation.

All RGGI states have adopted virtually uniform state rules to implement the RGGI program. The RGGI states decided to auction all of their allowances to the highest bidders rather than allocated to emission sources. For Massachusetts, this auction, even at only $5 per allowance, would raise more than $100 million annually.[9] The results of the March 2009 RGGI auction of allowances are shown in table 7-1.

Table 7-1. RGGI results of auction 3, March 18, 2009

State	CO_2 Allowances Auctioned	Clearing Price	2012 Vintage Proceeds	Cumulative Proceeds*
Connecticut	120,319	$3.05	$366,972.95	$14,037,371.75
Delaware	66,698	$3.05	$203,428.90	$5,439,723.34
Maine	65,437	$3.05	$199,582.85	$8,889,742.61
Maryland	399,884	$3.05	$1,219,646.20	$54,324,191.98
Massachusetts	328,565	$3.05	$1,002,123.25	$44,579,161.89
New Hampshire	86,850	$3.05	$264,892.50	$8,461,308.91
New Jersey	283,298	$3.05	$864,058.90	$32,094,782.19
New York	776,385	$3.05	$2,367,974.25	$87,956,663.54
Rhode Island	32,908	$3.05	$100,369.40	$4,470,561.95
Vermont	15,169	$3.05	$46,265.45	$2,060,794.97
Total	2,175,513		$6,635,314.65	$262,314,303.13*

* Includes proceeds from 2012 and 2009 vintage allowances

These states have realized that instead of allowing the value of freely allocated allowances to affect the price at which electricity ultimately is sold—thereby allowing power producers to keep any windfall—the state could capture the windfall by auctioning all of the allowances and simultaneously requiring that the proceeds of these auctions be directed toward self-determined public benefits. Generators can sell any excess allowances or purchase additional allowances from other qualifying power producers. A bidder can purchase up to a maximum of 25% of allowances at a given auction.[10] Since any speculator may bid for allowances, there is no guarantee that existing carbon-emitting electric power plants will be successful bidders, and they could be short of the necessary allowances to continue their operations.

Use of RGGI Carbon Offsets and Ineligibility of Renewables

The RGGI scheme also creates an offsets program. *Offsets* under RGGI are emissions reductions that come from sources other than the fossil fuel-fired electricity generators that are subject to the carbon emissions cap under RGGI. The offsets program awards offset allowances for approved offset projects that were realized on or after the date of the MOU.[11] Power producers can use offset allowances to comply with part of their carbon cap emission requirements.[12]

Importantly, and somewhat controversially, offsets cannot be created by the installation of renewable energy generation or resources. To some, this would seem to be counterintuitive and to conflict with other policies. For example, 27 states and the District of Columbia award renewable energy credits for the installation of eligible[13] renewable energy electric generation facilities[14] (see chapter 18). Sixteen states also authorize a tax on retail utility bills that creates a renewable energy trust fund used to make grants, loans, or otherwise provide incentives to renewable energy projects.[15] In addition, 80% of the states allow eligibly defined smaller renewable energy projects to enjoy the net metering of their electricity when sold back to the host electricity supplier, thus effectively allowing these entities to sell wholesale power at retail rates (see chapter 18).[16] In sum, renewable resources are significantly promoted by a majority of states as solutions to the GHGs and other pollutants created by traditional fossil fuel-fired sources.

However, no carbon offset credit is allowed for any project that has an electric generation component, unless the project sponsor transfers the legal rights to the credits to the regulatory agency.[17] Moreover, the RGGI Model Rule disallows offset allowances for any offset project that receives funding or other incentives from one of the 16 state renewable energy trust funds discussed previously, or any credits or allowances that would be earned from any other mandatory or voluntary GHG programs.

What qualifies as an RGGI carbon offset?

Most offsets eligible under the RGGI Model Rule are created by manipulating biomass or agricultural resources, for example, through afforestation and methane capture. Afforestation projects, unless insurance against biomass loss is purchased for the forest, receive offset credits equal only to 90% of their absorption of CO_2, to account for possible loss of forest mass over time. The RGGI program also depreciates any savings from forestation by 20% to account for the possibility of catastrophic losses. In addition, to ensure permanent forest use, a restrictive conservation easement is required for forest projects that create offset credits.[18] Thus, for some RGGI states, the in-state agriculture RGGI offset opportunities are minimal. (See chapter 20 for more on natural carbon sequestration and forestation.)

No RGGI carbon offset credits can be awarded for projects that are required by any local, state, or federal law, regulation, or administrative or judicial order.[19] Thus, retrofits of better technology, efficiency improvements, or emission reductions required by regulation or embodied in permits or consent decrees will not create salable offset credits. Furthermore, the MOU places limits on the use of offsets and the issuance of additional offsets to moderate offset price impacts.[20] RGGI initially allows offset projects sited anywhere in the United States if the average price of an emission allowance remains below $7 per ton. In each compliance period, each generator will be allowed to cover up to 3.3% of its emissions using offset allowances, which is roughly equal to half of that generator's emissions reduction obligation.

If allowance prices rise above $10 per ton, RGGI will allow sources to cover up to 10% of their carbon emissions with offsets, and will allow offset projects outside the United States as well as permit the transfer and application of allowances created under the EU Emissions Trading Scheme (EU-ETS) and the Kyoto Protocol's Clean Development Mechanism (CDM). Accordingly, "the compliance period will be extended by one year, for a maximum compliance period of four years." The decision to include EU-ETS and Kyoto CDM project credits as eligible offset "currency" is curious. Since EU-ETS credits in both the Phase I and Phase II periods are given away without charge by EU countries to their industries, this effectively works as an income and welfare shift from U.S. power generation owners to EU industries. If the United States were to link any of its CO_2 regulatory systems with the EU scheme, commentators note that it would increase the relative cost of U.S. allowances and compliance. But offsets for projects located in non-RGGI U.S. states may only be awarded if such states are implementing similar greenhouse gas budget/trading programs and enter into an agreement to ensure the credibility and validity of offset allowances from that state.

Offset credits that are created have a lifetime of 10 years, with the possibility of renewal; afforestation projects create credits with a 20-year lifetime, with a possible renewal up to 60 years.[21] The RGGI Model Rule indicates that when a regulated entity's emissions exceed its CO_2 allowance budget, the state can deduct from the entity's compliance account future allowances (beyond the current control period) equal to three times the number of the entity's excess emissions.[22]

Environmental groups attacked the Rhode Island RGGI plan, which will try to auction all carbon allowances, since this would make an exception to give certain allowances for free while others are to be auctioned. The groups argue that this would create a windfall for early reducing power plants that made reductions for reasons not related to RGGI and increase the state emissions cap by the amount of these additional allowances. Similar arguments were made by environmental groups in Maine and New York. New York would allow a December 20, 2005 through 2008 early reduction period, and the Pace Energy Center and other environmental groups have argued that any early reduction would not have been motivated by the RGGI program but by other factors; therefore, early reduction credits should reduce those allowances otherwise available under the program, rather than adding to it. This interpretation appears contrary to an RGGI guidance document. Pace argues that they are concerned about an "over-allocation of allowances," and the "burden of proof should rest with the emitter-applicant" to prove that its carbon reductions are "additional."

This *additionality* requirement and early compliance credits are controversial concepts (see chapter 15). First, early reduction credits that add to the total envelope of allowances are embedded as an option in the RGGI Model Rule that 10 states use as the template for their programs.[23] Early compliance allowances recognize CO_2 reductions at covered power projects. Early compliance also was incorporated in other U.S. credit trading programs, such as for SO_2.[24]

Second, in the RGGI scheme, the ownership and use of early or other allowances are not dependent on additionality. In contrast to allowances, RGGI does require some additionality in the creation of offsets, which are legally distinct from allowances, but directly increase the quantity of carbon emission rights for larger power plants.[25] Offsets must be real, verifiable, permanent, enforceable, and "additional." In fact, the amount of RGGI allowances is allocated to each RGGI state based on the cumulative total of historical emissions of all large power plants in the state. Thus they contain no concept of additionality compared to business-as-usual emissions; in fact, they *are* business-as-usual emissions with no additionality. RGGI allowances can be owned or traded by any party or speculator, including parties that contribute no additionality to any reduction of carbon. RGGI offset projects must be commenced and completed after December 20, 2005, while early compliance RGGI allowance projects must be commenced and completed prior to 2009.

While the RGGI program was the first and only U.S. cap-and-trade carbon control program in force prior to 2011, there are other important fairly well-formed pending initiatives. Chapter 8 turns to the even more comprehensive California carbon control regulation, which even though not due to be implemented and enforced until 2012, attracts much legal and policy attention. Chapter 9 focuses on the two regional programs in the West and Midwest of the United States involving 13 states, as well as voluntary carbon trading programs. Collectively, these various state programs make up the constellation of U.S. state carbon regulation in 23 states, which, alone, would be the largest participant in the Kyoto Protocol were it legally applicable.

Notes

1 Regional Greenhouse Gas Initiative. 2003. *Goals, Proposed Tasks, Short-Term Action Items*. 1. http://www.rggi.org/docs/actionplanfinal.pdf.

2 Regional Greenhouse Gas Initiative. 2005. *Memorandum of Understanding*. http://www.rggi.org/docs/mou_final_12_20_05.pdf.

3 Sussman, Edna. 2006. New York Addresses Climate Change with the First Mandatory U.S. Greenhouse Gas Program. *New York State Bar Journal*. (May): 43–44.

4 Bolster, Heddy. 2006. The Commerce Clause Meets Environmental Protection: The Compensatory Tax Doctrine as a Defense of Potential Regional Carbon Dioxide Regulation. *Boston College Law Review*. 47 (4): 737, 744 (citing RGGI MOU, 2).

5 Stavins, Robert. 2007. *Addressing Climate Change with a Comprehensive U.S. Cap-and-Trade System*. Faculty Working Paper Paper RWP07-53, Kennedy School of Government, Harvard University, 2, 3. http://ksgnotes1.harvard.edu/Research/wpaper.nsf/rwp/RWP07-053/$File/rwp_07_053_stavins.pdf.

6 Regional Greenhouse Gas Initiative. 2007. *Model Rule*. Sec. 20-5.3(a)-(b). http://www.rggi.org/docs/model_rule_corrected_1_5_07.pdf.

7 European Climate Change Programme. 2008. *National Allocation Plans: First Phase (2005–2007)*. http://ec.europa.eu/environment/climat/emission/emission_plans.htm.

8 U.S. Congress. Staff of House Committee on Energy and Commerce. 2008. *Climate Change Legislation Design White Paper: Appropriate Roles for Different Levels of Government*. 110th Cong. http://energycommerce.house.gov/Climate_Change/white%20paper%20st-lcl%20roles%20final%202-22.pdf.

9 Cash, Cathy and Paul Whitehead. 2008. All-Auction Allowances Get Promoted on Capitol Hill, as Europe Heads that Way. *Electric Utility Week.* January 28: 1.

10 Regional Greenhouse Gas Initiative. 2008. *Design Elements for Regional Allowance Auctions under the Regional Greenhouse Gas Initiative.* March 17: sec. 1, 1–2. http://rggi.org/docs/20080317auction.design.pdf.

11 Regional Greenhouse Gas Initiative. 2005. Memorandum of Understanding: 4.

12 Ibid. The initial offset projects that can be approved under the offsets program include: (1) landfill methane capture and combustion; (2) sulfur hexafluoride (SH_6) capture and recycling; (3) afforestation (transition of land from a non-forested to forested state); (4) end-use efficiency for natural gas, propane and heating oil; (5) methane capture from farming operations; and (6) projects to reduce fugitive methane emissions from natural gas transmission and distribution.

13 There is significant variation in what is an eligible renewable energy technology in each of the states: while certain wind and solar technologies seem to qualify everywhere, the eligibility of various biomass, landfill gas, hydroelectric, and other facilities varies significantly. See Ferrey, Steven. 2004. Sustainable Energy, Environmental Policy, and States' Rights: Discerning the Energy Future Through the Eye of the Dormant Commerce Clause. *Environmental Law Journal.* New York University. 12 (507): 646, table 3.

14 For a detailed discussion of these programs, see Ferrey, *Sustainable Energy, Environmental Policy, and States' Rights*, 529–532. See also, Ferrey, Steven. 2006. Renewable Orphans: Adopting Legal Renewable Standards at the State Level. *Electricity Journal.* (March): 52.

15 See Database of State Incentives for Renewables & Efficiency. 2007. *Rules, Regulations, & Policies for Renewable Energy*; Ferrey, *Sustainable Energy, Environmental Policy, and States' Rights*, 523; http://www.dsire.org.

16 For a discussion of net metering and its legal and policy implications, see Ferrey, Steven. 2003. Nothing But Net. *Duke Environmental Law & Policy Forum.* 14 (1); See also, Ferrey, Steven. 2004. Net Zero: Distributed Generation and FERC's MidAmerican Decision. *Electricity Journal.* (October): 33–34.

17 Regional Greenhouse Gas Initiative. 2007. *Model Rule.* Sec. 20-10.3(d)(2). http://www.rggi.org/docs/model_rule_corrected_1_5_07.pdf.

18 Ibid, sec. 10.5(c)(6)(i).

19 Ibid, sec. 10.3(d)(1).

20 Regional Greenhouse Gas Initiative. 2005. *Memorandum of Understanding in Brief.* http://www.rggi.org/docs/mou_brief_12_20_05.pdf.

21 Regional Greenhouse Gas Initiative. 2007. *Model Rule.* Sec. 20-10.3(e)(2). http://www.rggi.org/docs/model_rule_corrected_1_5_07.pdf.

22 Ibid, sec. 20-6.5(d)(1).

23 Regional Greenhouse Gas Initiative. 2007. *Model Rule.* Sec. 20-5.3. http://www.rggi.org/docs/model_rule_corrected_1_5_07.pdf.

24 Code of Federal Regulations. 40 sec. 73.71(a)-(f) (1991).
25 Regional Greenhouse Gas Initiative. 2007. *Model Rule.* sec. 20-10.3(d). http://www.rggi.org/docs/model_rule_corrected_1_5_07.pdf.

8) GOLDEN STATE CARBON: CALIFORNIA GHG REGULATION

California's Unique Approach to Carbon Regulation

California has an interesting power profile: it enjoys lower per capita energy consumption than the nation as a whole. However, approximately two-thirds of this lower amount is attributed to California's moderate climate and types of industry.

California faces its own challenge to control peak electric demand growth. Despite a meltdown of the restructured electric sector in 2000–2001, which caused the recall of the governor and the realization that electric demand had been allowed to outstrip supply additions, California's peak demand follows a pattern.[1] The state's peak electric load increased by 38% between its energy crisis in 2001 and 2006, or a notable increase in demand of more than 6% annually in peak demand.[2] This is about more than three times the national average rate of increase of less than 2% annually.[3]

The California carbon choices in AB 32

The California carbon scheme requires that California reduce GHG emissions to 1990 levels by 2020, counting all in-state and out-of-state generation used to serve California electric load.[4] California greenhouse gas emissions in 2004 were already almost 15% greater than in the 1990s. Pursuant to the California Global Warming Solutions Act of 2006 (commonly referred to as Assembly Bill 32 or AB 32), the state is required to reduce its aggregate GHG emissions to 1990 levels by 2020. This equates to an eventual estimated 25% reduction from business-as-usual levels.[5]

AB 32 charges the California Air Resources Board (CARB) with the responsibility of developing and implementing a plan to meet this challenging emissions-reduction goal. CARB is authorized, but not required, to establish and enforce a market-based compliance system, which could include carbon credits and banking. CARB has the responsibility of establishing a statewide GHG emissions cap for implementation in 2020 based on 1990 emissions levels by January 1, 2008. AB 32 further requires CARB to do the following:

- Adopt by January 1, 2008 regulations that require mandatory reporting and verification for significant GHG sources and to monitor compliance

- Adopt a plan by January 1, 2009 for achieving emissions reductions from significant GHG sources via regulations, market mechanisms, and other actions

- Adopt rules and regulations by January 1, 2011 to achieve the maximum technologically feasible and cost-effective GHG reductions, including provisions for using both market mechanisms and alternative compliance mechanisms

- Evaluate several factors—prior to imposing mandates or implementing market mechanisms—including but not limited to "impacts on California's economy, the environment, and public health; equity between regulated entities; electricity reliability; conformance with other environmental laws"; and evaluate whether the rules will disproportionately impact low-income communities

The regulatory deadlines under AB 32 in California are set forth in table 8–1.

Table 8–1. California Air Resources Board deadlines

Target Date	Task
2007	California Air Resources Board (CARB) must determine early action measures
2008	CARB must establish monitoring and reporting system
2008	CARB must determine 1990 GHG emissions levels
2009	CARB must have plan for meeting targets in the most cost-effective manner
2010	Meet deadline for implementing early action measure
2011	Regulations for meeting targets must be in place
2020	Affected entities begin to phase in GHG limits
2020	Reach 1990 emissions levels

The Advisory Committee changes

To assist CARB in fulfilling its charge, the governor created the Market Advisory Committee (MAC) to advise CARB on the development of a statewide plan to reduce GHG emissions. MAC's primary objective was to design a mandatory cap-and-trade program to achieve cost-effective emissions cuts across all sectors.[6] MAC recommends the following recommendations in its final report.

Expand cap-and-trade. California's cap-and-trade program should eventually incorporate all major GHG-emitting sectors in the state, with greatest attention to the electricity, industry, buildings, and transportation sectors, covering as many sectors, sources, and gases as possible, under mandatory reporting requirements.

Allowances. The initial scheme should include free allocation of some allowances and auctioning the other share of allowances, with the percentage of allowances auctioned increasing over time. MAC encourages the state to retain flexibility to freely allocate some of the allowances in a manner that stabilizes the price impacts and manages competitiveness among California power producers.

Offsets. The cap-and-trade program should recognize offsets generated by sources within and outside of California's borders.

Global GHG market. California's cap-and-trade program should be linked to similar policy initiatives in other jurisdictions to actively promote a "global greenhouse gas market."

First-seller approach. Because the quantity of California's imported electricity generated from coal is significant (56% of California's GHG emissions are from power imports from out-of-state producers, while 44% of GHG emissions are from in-state sources), California's cap-and-trade program should take a *first-seller* approach to capping emissions associated with electricity. Under this approach, the entity that first sells electricity within the state must meet the compliance obligation established under the cap-and-trade scheme. For power generated in California, the owner or operator of the in-state power plant is considered the first seller and would be required to meet the emissions cap. For imported power, the first seller is typically an investor-owned or municipal utility or wholesale power marketer that sells electricity to a load-serving entity or large end user. The out-of-state entity under this approach would also be required to meet the emissions cap.

This final MAC recommendation represents a significant departure from the original legislative scheme, and this departure is legally significant in evaluation of the scheme's legal nature. Originally, California intended to regulate GHGs from the utility sector by regulating all retail electric load-serving entities (LSEs), or retailers of power. Legally, all of these LSEs are located in-state or at least doing business in-state to be able to serve retail customers with electricity, and regulation would be imposed at the retail level on California activities. Note that regulators still regulate the allowed retail price of power sold by utilities and load-serving entities, if they end up being the entity that is directed to obtain carbon allowances.

However, this MAC recommendation, which in 2008 was accepted to be implemented by the Air Resources Board and Public Utilities Commission, shifts the point of control upstream to regulate power wholesalers at the first-seller transaction. This change makes it similar to the RGGI point of regulation. The restructuring of California's electric market in 1998 and the subsequent restructuring in 2001, due to an electric energy crisis, led to many changes.[7] Most of the power retailed in the state first goes through a wholesale power marketer as a wholesale transaction. Thus, many of these first sellers are now outside the state, and regulating first sellers causes the state at least indirectly to regulate the wholesale transaction, which traditionally is reserved to Federal Energy Regulatory Commission (FERC) authority at the federal level rather than to state authority. In addition, some of these transactions are across state lines, and many of the power transfers actually occur outside the state.

Legal distinctions

Even the variations within a state can make profound legal and policy differences. For example, there are almost 200 investor-owned utilities in the United States, which are regulated by public utility commissions in every state except Nebraska (which has no such private entities). In addition, there are approximately 2,000 municipal utilities, rural electric cooperatives, and public cooperative utilities. These typically are not regulated by the state regulatory commissions, although the scope of jurisdictional authority is different in every state. Of the 50 states, 18 deregulated retail electric power supply between 1997 and 2001. All of the other states were considering such restructuring and retail electric deregulation when the California electric market experienced a debacle in 2000–2001. No other state has since embarked on, or proceeded with, deregulation at the retail level.

However, that deregulation at the retail level, including in California and the RGGI states, did not include the unregulated municipal electric utilities. They were left to retain their retail monopolies and to remain vertically integrated. They were not forced to divest their generating assets, as were the investor-owned utilities subject to plenary state regulation. In California, these municipal utilities include the Los Angeles Department of Water and Power (LADWP), the largest municipal utility in the United States, serving 1.6 million customers and owning and operating more than 7,000 MW of power generation. Other significant municipal utilities in California include Sacramento, the state capitol; Anaheim, the home of Disneyland; and Palo Alto, the home of Stanford University and Silicon Valley.

Carbon regulation under AB 32 does not recognize these significant differences between municipal and investor-owned utilities in the state. All load-serving entities with retail power are subject to the same power sector carbon requirements. This same carbon requirement effectively, though, imposes a differential regulatory obligation on restructured private and traditional municipal utilities, as there are differences in ownership of carbon-producing resources. Some municipalities either own or have long-term contracts to purchase (and often from out-of-state generators engaging in interstate power transactions) high-carbon coal-generating power. Here, treating differently situated power retailing entities and LSEs the same for purposes of carbon regulation can yield very different legal and technical results.

Therefore, this decision on point of regulation for carbon control is more than just a cosmetic or policy choice. It has significant legal ramifications. While the MAC's accepted recommendation to the state would make the point of regulation parallel to that of RGGI, it raises legal issues, especially for a state like California where its LSEs, pursuant to long-term and legally binding contracts, import various forms of wholesale power from many other states. In fact, the Western Area Power Administration coordinates power transfers among 11 western states, including California. It has been noted by observers that California's choice to regulate carbon at the point of generation, or first delivery, is necessary for California to move upstream to get at the problem of high-carbon power leakage into the state.

However, any state must be careful with this direct or indirect external carbon regulation, as the federal Constitution prohibits a state regulating interstate commerce occurring in other states. If a state chooses to impose a fee or a tax on power moving in interstate commerce from another state, and keeps that revenue for its own purposes or citizens, in can substantially affect the flow of otherwise free interstate power. Power has been called by the U.S. Supreme Court the quintessential article in interstate commerce. It moves across state lines at almost the speed of light, based on instantaneous power demand factors not within government control. States are not allowed to exercise regulatory authority over commercial activities occurring outside the state. States cannot assess fees or taxes on such foreign transactions with the goal to benefit in-state ratepayers or in-state projects.

The California decision concluded that their system does not impose a license requirement on wholesale power transactions, and "the delivery point of regulation . . . does not regulate a commerce that occurs totally outside of California, and therefore does not regulate extraterritorially in violation of the Commerce Clause."[8] However, the Commerce Clause is broad. Moreover, general attributes are ascribed by California to power generated outside the state and owned by someone in the state. The California regulation attaches a condition to the very act of importing wholesale power into the state, which is not imposed by a typical state.

The debate as to where to monitor and assess CO_2 emissions is controversial. If regulation is imposed downstream in the flow of commerce, very different incentives are produced and different legal issues are raised than if regulation is imposed upstream. Certain industries, like mining and extraction, want to push the regulation away from their upstream operations to downstream users at power plants and in end-use applications. By contrast, end users of energy would prefer not to be the target of regulation.

LADWP criticized the recommendation of the California Public Utilities Commission (CPUC) to CARB to base allowance allocation on gross sales, irrespective of fuels used and generating mix, as taking funds from certain areas of the state and reallocating them to GHG-reduction efforts that may not benefit those who paid these amounts.[9] It also criticized the position of other investor-owned utilities to support the effect of a cap-and-trade allowance auction system to "true-up" the costs of more coal-sourced electric utility rates with those of other investor-owned utilities.[10]

The Los Angeles Department of Water and Power filed comments to the California energy regulatory agencies stating that there needed to be a rehearing on legal issues associated with carbon regulation. They wanted more debate on whether auction of allowances does the following:[11]

- Creates a tax in violation of California Proposition 13
- Is differentially applied to utilities as compared to other businesses in violation of the state constitution
- Violates home rule authority of the City pursuant to state law

- Unconstitutionally transfers funds from utility providers, by means of a forced regulatory gift

Under the auction scheme, utilities that used their ratepayers' monies to acquire needed allowances at auction would not see those funds devoted to projects that directly credited or benefited those ratepayers.[12]

Specific California Carbon Emission Standards

The California scheme covers all load-serving entities, including municipal LSEs. Electric generators are required to meet a CO_2 emissions level no higher than that achievable by a combined-cycle gas-fired power generator.[13] Any new contracts for a term of five years or more for the procurement of baseload generation must comply with a performance standard of emitting no more than 1100 lbs CO_2/MWh of power generation.[14] *Baseload* generation is defined as generation that is designed and intended to operate at an annualized capacity factor of 60% or greater.

Additionally, AB 32 specifically requires the California Air Resources Board to consider cumulative impact of direct and indirect sources of emissions on adversely affected communities. Environmental groups have argued that the program must reduce emissions within California, and that it must not permit the use of offsets that are purchased from out of the state.

California distinctions: coal

California itself has few coal-fired plants, relying in-state for generation 50% from natural gas, 20% from hydroelectric power, and 16% from nuclear power. Less than 2% of in-state California power is fired by coal. These percentages of lower-carbon power generation in-state are more than twice those of the nation as a whole. However, significant amounts of California power comes from outside the state, and roughly half of California's electric sector GHG emissions are the result of electric power imports from out of state that are generated predominately by coal-fired power plants.[15] This imported power into California constitutes only about a quarter of the total California consumption, but half of the carbon emissions.

The impact of California's new emissions limitations will thus significantly reduce the attractiveness and viability of coal-fired generation for California. While California has little in-state coal generation, various California LSEs, particularly the Los Angeles Department of Water and Power, import significant amounts of coal-fired power from various other states. The LADWP has argued that it serves a lower-income population and that the change to a first-seller point of CO_2 regulation is unfair treatment and targets Southern California.

In a meeting held by the California Public Utilities Commission and the California Energy Commission to review various proposals for AB 32 implementation, Southern California Edison proposed to have coal-fired generators to be given free allowances to shield ratepayers from carbon allowance costs.[16] The LADWP requested an opt-out option from the cap-and-trade requirements. The utility stated that if it had to comply with California's carbon cap-and-trade requirements, it would have to either jettison its renewable energy program or raise rates substantially. It argued, instead, that if a utility system implements a vigorous reduction of carbon emissions through renewable energy programs, that should be the metric for carbon compliance and reduction, rather than penalizing it for its long-term power contracts that obligate it to purchase higher carbon resources as a matter of contract law. For example, LADWP has coal-unit contracts, out of state, to purchase power until 2019 and 2027.

There also can be other differences in certain California carbon configurations. While most areas of the United States have electric power sector carbon emissions leading transportation sector emissions, this is not true in California. In California, transportation accounts for more than twice the level of carbon emissions as the power sector. However, the transportation sector is not the prime target of California carbon regulation. In addition, some areas of California are quite post-industrial, with a relatively small industrial base, and a larger commercial base. This can alter the onsite power demand reduction options. In California, where heating and cooking fuels are supplied predominantly by natural gas, these carbon-related emissions can be significant and almost as large as power sector emissions. However, California's carbon requirements on LSEs do not initially include accounting for these carbon emissions from natural gas distributed by retail utility companies.

California distinctions: LSEs

In addition, there are key functions that certain cities serve. For example, the Ports of Los Angeles and Long Beach bring in more than 40% of all the ship container cargo in the United States. They are very energy-intensive in electricity, truck transport traffic, and use of fossil fuels while in port. There are also four airports in Los Angeles County, including one of the busiest in the world. These sectors of the economy provide a commercial or industrial function for a wider area than just the city, and the associated carbon footprint from such activities is a much wider geographic imprint than just services for the city itself. How these historic carbon emissions are factored into carbon reduction requirements becomes a critical factor.

Pursuant to AB 32, utilities are required to "account for greenhouse gas emissions from . . . electricity generated within the state or imported from outside the state."[17] The California scheme thus impacts all in-state and out-of-state generation used to serve California's electric load. It does not distinguish the geographic source of power generation, and covers the liberal flow of power into California from other states.

California's carbon regulation system was originally intended to be different from that of RGGI in that the carbon compliance obligation of the former was to be placed on load-serving entities, rather than generators of power. This is a distinction of whether carbon regulation covers the generator of the power or the distributor of the power. Load-serving entities are distributors of retail power, such as utilities or retail suppliers. LSEs have an entire portfolio of power generation resources that they can optimize and blend for purposes of carbon limitation compliance. They can continue to purchase carbon-rich generation, and compensate by adding renewable energy resources or other low-carbon generation to achieve the average requirements over their entire portfolio of power generation resources.

If the California MAC recommendations are followed, however, both California and RGGI schemes would regulate at the individual wholesale generator level.[18] Like the RGGI scheme, the MAC recommendations would require each and every power generator or first seller of power to comply individually, penalize high-carbon generating resources *per se*, and not allow any optimization within portfolios of generation. In RGGI and the revised California protocol of regulation, each individual power generation facility is responsible for compliance.

California issues: challenges

The RGGI and California systems, while similar, are not identical. RGGI regulates only CO_2 and regulates only the electric power sector, and then only larger plants in that sector. California regulates all GHGs, including CO_2. RGGI began in 2009; California's carbon program begins in 2012. This delay may be California's hope regarding legal challenges—perhaps federal requirements will begin before the 2012 commencement date, preempting state regulation, and thus mitigating the constitutional uncertainty.

The constitutional issues confronting these state carbon regulation efforts are that the agreement and the means of its implementation may or may not violate the Supremacy Clause and the Compact Clause of the U.S. Constitution (see chapters 13–14). Moreover, in order for RGGI to work effectively at reducing carbon emissions, rather than increasing the importation of carbon-intensive electricity, some states are actively considering surcharging or taxing wholesale power leaking in from outside the region, which itself might violate the Commerce Clause. In certain states, the carbon scheme has been accused of constituting an unauthorized tax because most states require taxes to be created by legislatively-passed statutes, not regulations enacted by the executive branch. Some states also impose limitations on new taxes or supermajority votes to authorize them. The debate then evolves to whether the carbon allowance auction and cost of acquisition constitutes a tax or a fee. Even if deemed to constitute a fee, the revenues gained from the auction can far exceed the costs associated with the program. This can raise legal questions about the validity of a fee for general revenue raising, unrelated to the specific program. Such large fees have been characterized by some courts as equivalent to taxes.

These legal controversies are examined in detail in chapters 13–14. However, the next chapter looks at regional and voluntary carbon regulation initiatives in the United States.

Notes

1 Ferrey, Steven. 2004. Soft Paths, Hard Choices: Environmental Lessons in the Aftermath of California's Electric Deregulation Debacle. *Virginia Environmental Law Journal.* 23 (2): 251, 255.

2 See, Hilton, Seth. 2006. The Impact of California's Global Warming Legislation on the Electric Utility Industry. *Electricity Journal.* Vol. 10(9): 10, 11. During this period, California increased its generation supply by only 23% since the 2001 crisis.

3 See, Energy Information Administration. 2008. *Annual Energy Review.* Department of Energy. Table 8.2(a). http://www.eia.doe.gov/aer/pdf/pages/sec8_8.pdf.

4 California Global Warming Solutions Act of 2006. 2007. California Health and Safety Code. Sec. 38,500–38,599.

5 California Energy Commission. 2008. *History of California's Involvement in Air Pollution and Global Climate Change.* http://www.climatechange.ca.gov/background/history.html.

6 Market Advisory Commission to the California Air Resources Board. 2007. *Recommendations for Designing a Greenhouse Gas Cap-and-Trade System for California*. http://www.climatechange.ca.gov/events/2007-06-12_mac_meeting/2007-06-01_MAC_DRAFT_REPORT.PDF.

7 Ferrey, *Soft Paths, Hard Choices*, 297.

8 California Public Utilities Commission. Proposed Decision. 71–72; Los Angeles Department of Public Works. 2008. *Opening Comments of LADWP on the Proposed Decision of President Peevey: Interim Opinion on Greenhouse Gas Regulatory Strategies*. February 28: 13.

9 Los Angeles Department of Public Works. 2008. *Reply Comments on Proposed Decision of Commissioner Peevey: Final Opinion on Greenhouse Gas Regulatory Strategies*. October 7: 3.

10 Los Angeles Department of Public Works. *Opening Comments of LADWP on the Proposed Decision of President Peevey: Interim Opinion*, 8.

11 Los Angeles Department of Public Works. 2008. *Request for Rehearing/Reconsideration of LADWP on Final Opinion on Greenhouse Gas Regulatory Strategies*. November 21: 5.

12 Los Angeles Department of Public Works. *Opening Comments of LADWP on the Proposed Decision of President Peevey: Interim Opinion*, 8.

13 California Public Utilities. 2007. Code Sec. 8341(d)(1). This legislation targets only electric generation. Sections 8340 and 8341 govern all new long-term energy commitments and establish a "greenhouse gas emissions performance standard". Ibid, sec. 8340–8341. This is specific to the electric power role in meeting AB 32 goals. The GHG emissions standard creates a specific level of permissible emissions and prohibits new construction, new long-term power contracts, and any major plant investment that will not meet the performance standard. This prohibits load-serving entities from entering long-term power contracts with out-of-state producers who do not meet California's stringent new emissions standard. California's Public Utilities Commission (PUC) has set the GHG emissions performance standard at the equivalent of the emissions from a combined-cycle natural gas plant.

14 Ibid, sec. 8341(b), (d)(1). This is a level that conventional coal-fired electric generation will not be able to meet, generating about 1,770 lbs. CO_2/MWh.

15 California Public Utilities Commission. 2007. *Order Instituting Rulemaking to Implement the Commission's Procurement Incentive Framework and to Examine the Integration of Greenhouse Gas Emissions Standards into Procurement Policies*. R.06-04-009, D.07-09-017, 2007 Cal. P.U.C. LEXIS 330. Three-quarters of California's power imports come from the Southwest, and involves much coal-fired power, as opposed to the other quarter that is imported through the Northwest. Alvarado, Al and Karen Griffin. 2007. *Revised Methodology to Estimate the Generation Resource Mix of California Electricity Imports*. California Energy Commission update to the May 2006 Staff Paper 1, 1, 3.

16 Weinzimer, Lisa. 2008. Debate Heats up over Allocating CO_2 Allowances in Calif., Generators Deny Windfall Is Possible. *Electric Utility Week*. April 28: 13. Load

served and historical emissions both would be factors in determining the amount of allowances given. Ibid.

17 California Health and Safety. 2007. Code. Sec. 38530(b)(2). "This requirement applies to all retail sellers of electricity, including load-serving entities as defined in subdivision (j) of Section 380 of the Public Utilities Code and local publicly owned electric utilities as defined in Section 9604 of the Public Utilities Code." Ibid.

18 The RGGI system governs only the original power producers, whereas the California bill governs any load-serving entity, defined as "every electrical corporation, electric service provider, or community choice aggregator serving end-use customers in the state." California Public Utilities. 2007. Code Sec. 8340(h).

9. REGIONAL AND VOLUNTARY U.S. CARBON PROGRAMS

Voluntary Domestic Carbon Regulation

The United States has some national voluntary carbon programs. With the passage of the Energy Policy Act of 1992, Congress authorized a voluntary program to encourage the public to report achievements in reducing GHG emissions. Beginning in October of 1994, the U.S. Department of Energy issued guidelines on the voluntary reporting of emissions reductions and carbon sequestration.[1] This program, though, only offers an opportunity to report annual GHG emissions and record projects that reduce emissions or increase carbon sequestration, but it does not provide a mechanism or monetary incentives to reduce carbon emissions.

The Chicago Climate Exchange (CCX) was among the first to create a voluntary, legally binding multi-sector reduction and trade program that provides true monetary incentives. CCX is currently the single voluntary emissions trading system for all six GHGs and has almost 300 members from various sectors worldwide.[2] For CCX members who choose to voluntarily participate in CCX's binding commitment to meet annual GHG emission reduction goals, the program provides an opportunity to capitalize on the burgeoning carbon market.

CCX issues Carbon Financial Instrument (CFI) contracts, each representing the equivalent of 100 metric tons of CO_2, as the tradable commodity. The CFI contracts are either *Exchange Allowances* based on a member's emission baseline and an overall reduction schedule, or *Exchange Offsets* generated by certain types of offset projects. CCX members that reduce emissions below the target levels can sell or bank their surplus allowances.

Participation in the trading system requires that members agree to surrender their CFI contracts to meet the emission reduction requirement. To meet the requirement, members follow a schedule for reducing emissions that is carried out in two phases. Phase I (2003-2006) required members to commit to reduce 1% below the 1998-2001 baseline. This would result in a total reduction of 4% by 2006. Phase II (2007-2010) requires members to commit to an annual reduction schedule of an additional 2%, resulting in an overall reduction commitment of 6% below baseline.

The CCX's Offsets Program allows members and other entities that do not have significant GHG emissions to register offset projects.[3] CCX will issue traditional CFI contracts to Offset Providers or Offset Aggregators "for eligible projects on the basis of sequestration, destruction or displacement of GHG emissions." An Offset Provider is defined as an owner of an offset project that registers and sells offsets on its own behalf. An Offset Aggregator is defined as an entity that serves as the administrative representative, on behalf of offset project owners, of multiple offset-generating projects. Offset Aggregators register and sell offset projects involving less than 10,000 metric tons of CO_2 equivalent per year.

CFI contracts are issued by CCX according to standardized rules for projects involving agricultural methane, landfill methane, agricultural soil carbon, forestry, renewable energy, coal mine methane, and rangeland soil carbon. Other types of projects, such as energy efficiency and fuel switching, are approved by CCX on a project-by-project basis. Trading of carbon credits is already robust, and growing quickly. More than 600 separate entities develop, market, or sell offsets in the United States in markets that have limited transparency. The CCX uses a registry to track offset trades, which can occur across international borders through Web sites.

The national U.S. carbon regulation may recognize prior voluntary carbon reductions that have been registered and independently verified. This could create monetary value for prior carbon credits. Figure 9-1 provides a depiction of several proposed U.S. federal climate change legislation requirements over time.

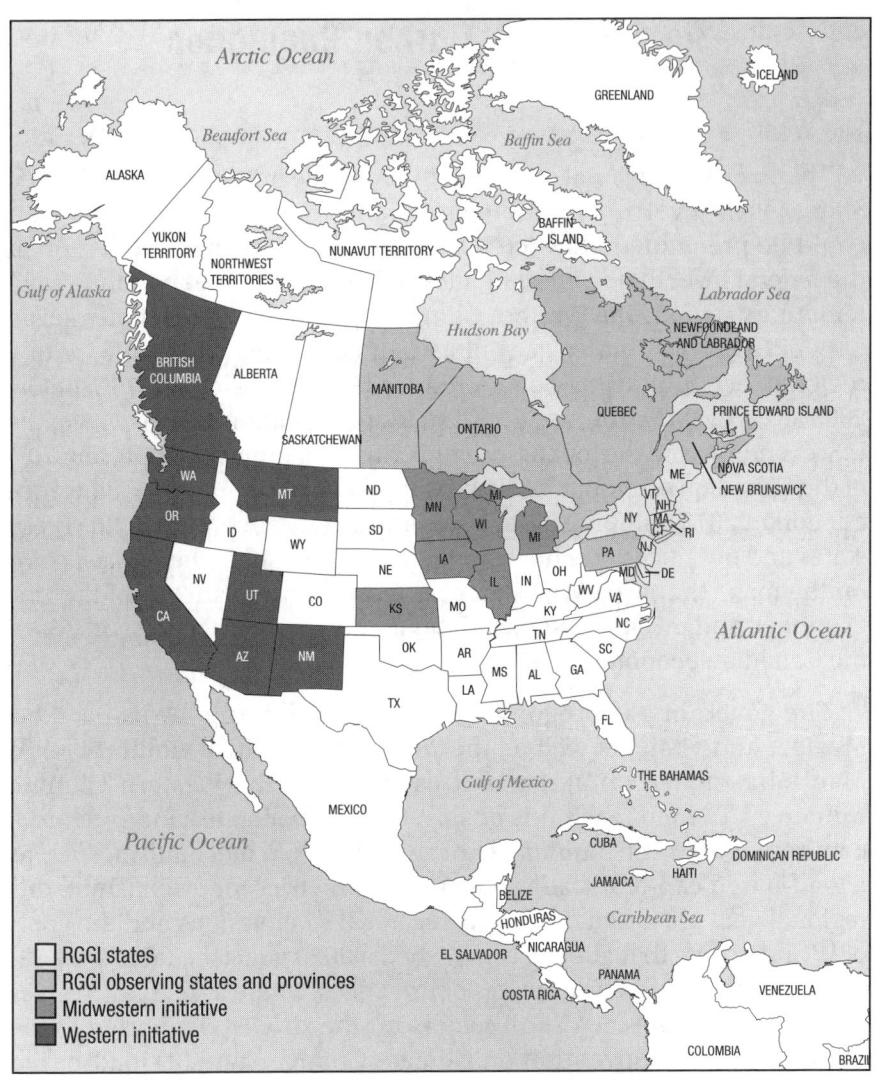

CHAPTER 9 • REGIONAL AND VOLUNTARY U.S. CARBON PROGRAMS   105

Fig. 9–1. State and regional climate initiatives

Regional U.S. Carbon Regulation

The West

There are two regional carbon cap-and-trade initiatives in the United States, each involving multiple states and Canadian provinces. They could be preempted between 2012 and 2017 by carbon regulation at the federal level that would not allow state regulation to continue for a temporary period. The Western Climate Initiative (WCI) includes seven very different U.S. states as well as four provinces in Canada, with a regional, economy-wide goal to reduce GHG emissions to 15% below 2005 levels by 2020. The seven WCI states represent more than 20% of the U.S. economy, and the four associated Canadian provinces represent 70% of the Canadian economy. WCI will start with a minimum 10% allowance auction in 2012, ramping up to a minimum 25% auction by 2020. These states and provinces are shown in figure 9-1 shaded in the darkest color on the map. The seven WCI states represent more than 20% of the U.S. economy, and the four associated Canadian provinces represent 70% of the Canadian economy.

The governors of Oregon, Washington, California, Arizona, New Mexico, and Utah, as well as the premiers of British Columbia and Manitoba, signed an agreement establishing the Western Climate Initiative.[4] To help reach this goal, WCI member states and provinces committed to unveil a multi-sector market-based mechanism, such as a load-based cap-and-trade program. The metrics for establishing this regional goal are based on: (1) aggregate GHG emissions and the goals of WCI partners that already established a 2020 goal (see table 9-1); (2) emissions inventories from states or provinces, where available; (3) gross emissions estimates across all sectors for the six GHGs reported to the UN Framework Convention on Climate Change;[5] and (4) load-based emissions estimates for the electricity sector.[6]

Table 9-1. State and provincial goals for GHG reductions

	Short Term (2010–12)	Medium Term (2020)	Long Term (2040-50)
Arizona	not established	2000 levels by 2020	50% below 2000 by 2040
British Columbia	not established	33% below 2007 by 2020	not established
California	2000 levels by 2010	1990 levels by 2020	80% below 1990 by 2050
Manitoba	6% below 1990	6% below 1990	not established
New Mexico	2000 levels by 2012	10% below 2000 by 2020	75% below 2000 by 2050
Oregon	arrest emissions growth	10% below 1990 by 2020	>75% below 1990 by 2050
Utah	....	Will set goals by June 2008	....
Washington	not established	1990 levels by 2020	50% below 1990 by 2050

New entrants making comparable efforts to reduce GHG emissions are encouraged to join the WCI. WCI members consider several factors when determining whether to admit a new state or province. Such factors include whether the proposed entrant has adopted an economy-wide reduction goal and developed a comprehensive plan to reach that goal, has agreed to adopt GHG tailpipe standards for passenger vehicles, and is participating in the Climate Registry.

To achieve the new regional GHG emissions reduction goal, WCI is committed to limiting emissions that contribute to climate change from all sources of GHGs, including but not limited to stationary sources, energy supply, residential, commercial, industrial, transportation, waste management, agriculture, and forestry. Eventually WCI's plan to curb emissions will focus on power plants and vehicles. Implementing the WCI plan will likely restrict the continued development of coal-fired power generation facilities because it will otherwise be difficult to meet the emission reduction goals. In developing its market approach, WCI members are engaging in discussions with leaders in the Regional Greenhouse Gas Initiative and may consider some variety of incentives, standards, and regulations similar to the approach California has taken to combat climate change.

The regulated emissions include across all sectors the six GHGs reported to the UN Framework Convention on Climate Change. Half of the WCI states have not been able to approve the necessary state legislation. The Western state WCI program will allow participating states to use CDM and JI Kyoto credits as offsets. CDM offset credits are created exclusively in developing countries under the Kyoto Protocol. Environmental groups have complained about the out-of-region geographic location of such offsets. Groups in WCI states are concerned that the GHG reduction plan there might "crush the market for RECs."[7] The concern is that the area will not be able to create more green power. They want the purchase of renewable energy credits (RECs) to be tied into the reduction of the GHG cap that will be imposed.

California is the lead state in forming the WCI, but now complains that it is not treated fairly in the emerging WCI legal construct. California complained that the WCI will impose an inordinate burden on the California power sector starting in 2012, by excluding restriction on the transportation sector until 2015. Because California utilities rely on out-of-state electricity imports, California utilities argue that they require extra allocation of any allowances if it is to be the early focus. They also urge the WCI to increase the currently considered 10% limit on the use of offsets to demonstrate compliance, to a higher value to allow California industry to comply with more cost-effective actions external to regulated power generation facilities. Environmental groups have countered that 10% is too high.

The Midwest

Shifting to the center of the nation, in November 2007, six participating Midwestern states and two Canadian provinces executed a regional greenhouse gas emission reduction strategy.[8] This included Minnesota, Illinois, Indiana, Iowa, Michigan, Kansas, Ohio, South Dakota, Wisconsin, Manitoba, and Ontario. Three of these nine states are observing rather than participating initially. On figure 9-1, the Midwestern states are shaded in the second-darkest color on the map. Figure 9-1 also shows the RGGI states and the six observing Canadian provinces, and Pennsylvania as an observing state that also constitutes the geographic corridor for so-called "leakage" of carbon in to the RGGI area. The group worked to develop a cap-and-trade carbon program in 2008 for implementation in 2010. This accord will not set a specific target but will attempt to cut emissions by 2020.

Recommendations would allow 10–50% of reductions to be achieved through use of offsets. There is dispute as to whether allowances can come from other states. The Midwestern Greenhouse Gas Reduction Accord will establish a system to enable tracking, management, and crediting for entities that reduce GHGs. This region depends heavily on coal-fired electric generation, and is therefore distinct technologically from both California and the RGGI states.

The RGGI, Western states, and Midwest states carbon regulation schemes collectively include about half of the United States plus Canadian provinces. RGGI only affects CO_2 from larger power plants, while the regional climate initiatives are looking at GHGs more broadly from various economic sectors. Therefore, the activities of this half of the states are the leading edge of existing and planned U.S. carbon regulation. They may be preempted by national carbon regulation in the United States.

Previous chapters have now explored world, EU, and U.S. carbon regulation in this section of the book. Now the third section of this toolkit will turn to exploring in detail the legal and regulatory issues surrounding carbon regulation. This is the first book to look at the pivotal legal issues of all of the following touchstone elements of climate change:

- What has gone wrong with the Kyoto Protocol, and how it must be reformed to have any possibility for succession
- The new green grid and its implications
- The major issue of *leakage* and the U.S. constitutional Commerce Clause
- Carbon allowance auction of allowances and the U.S. constitutional Supremacy Clause and legal preemption
- The requirement for legal *additionality* for all carbon offset credits required universally by the Kyoto Protocol, the EU-ETS, and RGGI

Notes

1. U.S. Department of Energy. 1994. Voluntary Reporting of Greenhouse Gas Emissions under Section 1605(b) of the Energy Policy Act of 1992: General Guidelines. http://www.eia.doe.gov/oiaf/1605/1605b.html.

2. Chicago Climate Exchange. *History.* http://www.chicagoclimatex.com/content.jsf?id=1.

3. Chicago Climate Exchange. 2007. *CCX Offsets Program.* http://www.chicagoclimatex.com/content.jsf?id=23.

4. The original agreement was signed in February 2007 by governors of Arizona, California, New Mexico, Oregon, and Washington. In May 2007, the state of Utah and the Canadian provinces of British Columbia and Manitoba joined WCI. The states of Kansas, Colorado, Wyoming, and Nevada; the Canadian provinces of Ontario, Quebec, and Saskatchewan; and one Mexican state, Sonora, will participate in WCI as observers. See, Cash, Cathy. 2007. Western Region Plan to Reduce GHG Emissions has Energy Suppliers Waiting for Specifics. *Electric Utility Week.* August 27: 20.

5. These six GHGs include carbon dioxide (CO_2), methane (CH_4), nitrous oxide (N_2O), hydrofluorocarbons (HFCs), perfluorocarbons (PFCs), and sulfur hexafluoride (SF_6).

6. Western Climate Initiative. 2007. *Statement of Regional Goal* (Attachment A: Metrics Used to Establish WCI Regional Goal). August 22: 3. http://www.westernclimateinitiative.org/ewebeditpro/items/O104F13006.pdf.

7. Carbon Control News. 2008. Critics Say Western GHG Plan Would Crush Renewable Trading Market. *Carboncontrolnews.com.* August 25: 1, 8.

8. See http://www.Midwesterngovernors.org.

PART III

THE LEGAL AND POLICY ISSUES CONFRONTING CARBON CONTROL WORLDWIDE:
MANIPULATING THE TOOLBOX OF REGULATORY OPTIONS

10. THE KYOTO CRITIQUE: THE URGENCY OF INTERNATIONAL REDESIGN

Monitoring Achievement on Carbon

Has Kyoto worked? Yes and no. It depends on how the question is phrased. The Kyoto Protocol has mobilized 38 countries of the world to start limiting their cumulative carbon emissions. This is a solid and creditable start. However, this start shows reductions on paper, but has not translated into perceptible reductions in carbon in either these regulated developed countries or certainly, in more than 160 developing countries for which Kyoto provides no requirements. In fact, carbon reductions may have occurred as much from high prices of fossil fuels and dramatic worldwide economic recession, as in response to regulatory programs on climate change.

If the question is asked whether the Kyoto Protocol will result in significant carbon reductions by the forecast *tipping point* of 2015, the answer certainly is "no" (see chapter 4). The Kyoto Protocol needs to be extended or reauthorized to even endure in any format past its scheduled 2012 end. In fact, some "smart money" is even betting that by the presumed 2015 tipping point for world climate impacts there will not even be in place an enforceable regulatory mechanism to make significant reduction of carbon emissions, let alone having achieved them by this fast-approaching date.

The World Wildlife Fund even accused the European Union of "playing tricks on the atmosphere" with the EU-ETS carbon control program. WWF claimed that at best, only a 4–5% reduction from 2009 carbon emission levels onward will be achieved in the European Union, despite taking credit for a goal of 20% reduction by 2020 (see chapter 6). The Kyoto Protocol is very unlikely to achieve its target of getting an average of 7% below 1990 GHG emission levels by 2012.

Between 1990 and 2004, the Annex I developed nations, excluding the countries with "economies in transition" (the former Soviet economies), increased GHG annual emissions by 12.1%. These developed countries were responsible for 18.6 billion tons of GHGs emitted annually. Of the developing nations, 122 reported 11.7 billion tons of GHG emissions in 2004. Therefore, approximately 40% of GHGs are from developing countries. This may actually understate the percentage because only 122 of about 160 developing nations are included in UN databases, and there may be data gaps and underreporting in some of the 122 countries that do report.

Let's Do the Math

Assuming that the Kyoto targets must be achieved by 2020, a world reduction in carbon is only achieved if the *developing* nations of the world don't base their increasing electrification predominantly on carbon-based fuels. If the more than three dozen developed Annex I nations potentially regulated by the Kyoto Protocol were to reduce their emissions by 20% from current levels, they would approximately achieve their Kyoto Protocol targets. Since these Annex I countries currently emit approximately 60% of world carbon, this would constitute an approximately 12% reduction in total world carbon emissions. Even a 12% world reduction, assuming other non-regulated country carbon emissions are static, does not hit the goal of 20% world reduction. It is also important to remember that the balance is changing quickly over time, with the share of GHGs from developed countries declining to a smaller percentage of the whole each year, and projected soon to become the minority share.

However, if the non-Annex I developing nations, representing about 40% of world carbon emissions now and growing, increase their electricity demand by the forecast 4% annually between 2007 and 2020, that would constitute a cumulative compounded increase of about 60% of power use in that 40% share of the total emissions propagated by developing countries. This does not even include their other sources of GHG emissions, such as in the transportation sector. If those developing nations utilize predominately fossil fuels for this power sector expansion (and transportation fuels), that is an increase of 24% in world carbon emissions due just to increased power consumption in developing nations. This increase is double the optimistic decrease of power sector carbon in developed nations. And it is important to note that there is no assurance that emissions in fact will even decrease in developed nations. Figure 10-1 illustrates the growing percentage of GHGs from developing nations.

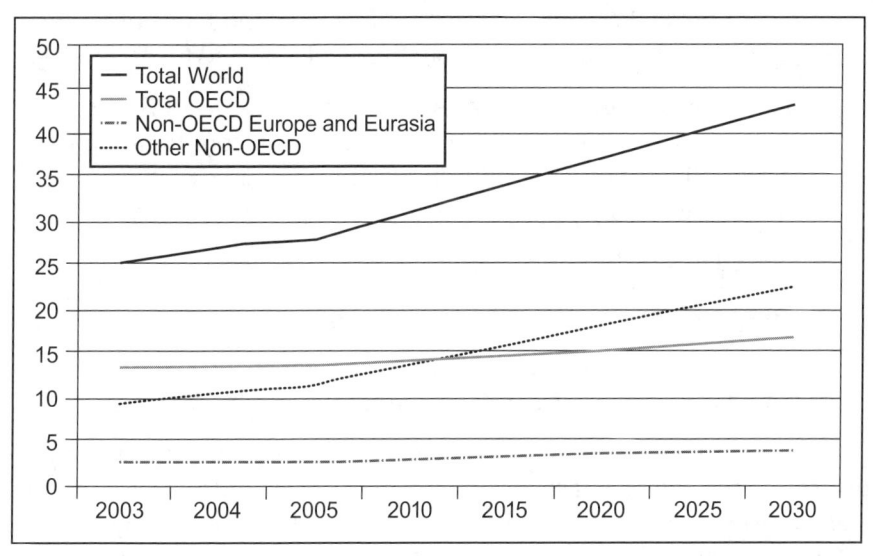

Fig. 10–1. World carbon dioxide emissions by region, 2003–2030 (billion metric tons of carbon dioxide)

Thus, the carbon increase in developing nations could totally negate, by a factor of two, the carbon reductions that the Kyoto Protocol seeks to achieve in Annex I developed countries, assuming that they are achieved. The use of fossil fuels for power generation in developing countries often employs less advanced combustion technologies that do not utilize the most effective emission control technologies. Moreover, it is important to note that the Kyoto Protocol is not yet now achieving its targets in the Annex I countries, with a composite 12% increase since 1990, so assumptions of success in the developed countries are unsure and could be optimistic compared to achievements.

Clearly, the Kyoto Protocol will have to be significantly redesigned, not just extended, to have any stake for effectiveness on the world stage in whatever time frame. The following section analyzes several dimensions of fundamental redesign of Kyoto that must be undertaken to make it an effective international mechanism to meet the challenges of global warming mitigation.

Fundamental Changes Required in the Protocol

Recent developments have not changed the Kyoto Protocol much. The December 2007 compromise guidelines from the Bali Conference of parties on continuation of the Kyoto Protocol changed the following:

- Backed off specific targets for Annex I countries to cut GHGs by 2020, and instead generally referenced IPCC recommendations of a 25-40% cut that the IPCC indicated is required to curb runaway warming
- Backed off any binding commitments
- Backed off any requirements on developing countries
- Set in motion the Bali Road Map of two years of discussions aimed at cutting GHGs by 2050, rather than by 2020

The Copenhagen conference in December 2009 ended in disjointed disarray. Thus, the recent set of agreements of the Kyoto parties does not stake either an aggressive or inclusive course. Goals and targets exist rather than requirements. Those goals have been stretched further out in time, at precisely the time when the scientific consensus is indicated a need for more urgency. The great majority of the nations of the planet are not covered by any Kyoto goals, targets, or requirements. No binding requirements have been imposed. No compulsion or enforcement mechanism exists. Nothing at the December 2009 Copenhagen conference changed this basic architecture, and only widened the Protocol's chasms.

The following section highlights three essential reforms that must be made to the Kyoto Protocol to alter it from an inaccurately drawn roadmap, to an effective international protocol or treaty. Emphasis is placed on what must happen to affect the international power sector, both in developed and developing countries. The success of addressing the power sector with these alterations will determine whether the Kyoto Protocol emerges as an effective international legal and regulatory instrument, or becomes a historical footnote to planetary climate change.

Reform 1: Enlarging the group of participating regulated countries

The new math of effective Kyoto architecture. This section starts with the first suggestion of change of Kyoto international carbon regulation. The quantitative goals in the Kyoto Protocol are not the fundamental problem—it is their application to only 25% of the nations that developed and ratified the Protocol. About three dozen countries are covered with Kyoto requirements, while about 160 countries, including 120 who ratified the Protocol, have no GHG reductions imposed under the Kyoto Protocol. Numerators are critical in the equation of Kyoto. The denominator of participating carbon contributors is a known operand of 162 world countries recognized in the United Nations. The numerator of the Kyoto Protocol is the critical operand to determine the participation of emitters. However, that numerator within the Kyoto Protocol only includes 38 Annex I countries, and in this fact lies a major shortcoming of the Protocol as an ineffective legal instrument on an international scale.

In 1997 the Kyoto Protocol assigned to each Annex I country a quantity of GHG emissions for the period 2008-2012. Developing nations successfully resisted efforts to include them in binding obligations and opposed encouraging their voluntary commitments to GHG reduction.[1] Kyoto reflects "common but differentiated" responsibilities between developed and developing countries.[2] That resistance continues to date.

The largest CO_2 emitter in the world, China, is not covered as an Annex I country. According to former British prime minister Tony Blair, "the vast majority of new power stations in China and India will be coal-fired; not 'may be coal-fired', will be."[3] In 2007, China built more new coal-fired power plants than Britain, the seat of the coal-fired industrial revolution, built in its entire history. This creates economic advantages and *leakage* of greater emissions from countries not covered by the Kyoto Protocol (see chapter 13). The share of GHG emissions by fuel are shown in figure 10-2.

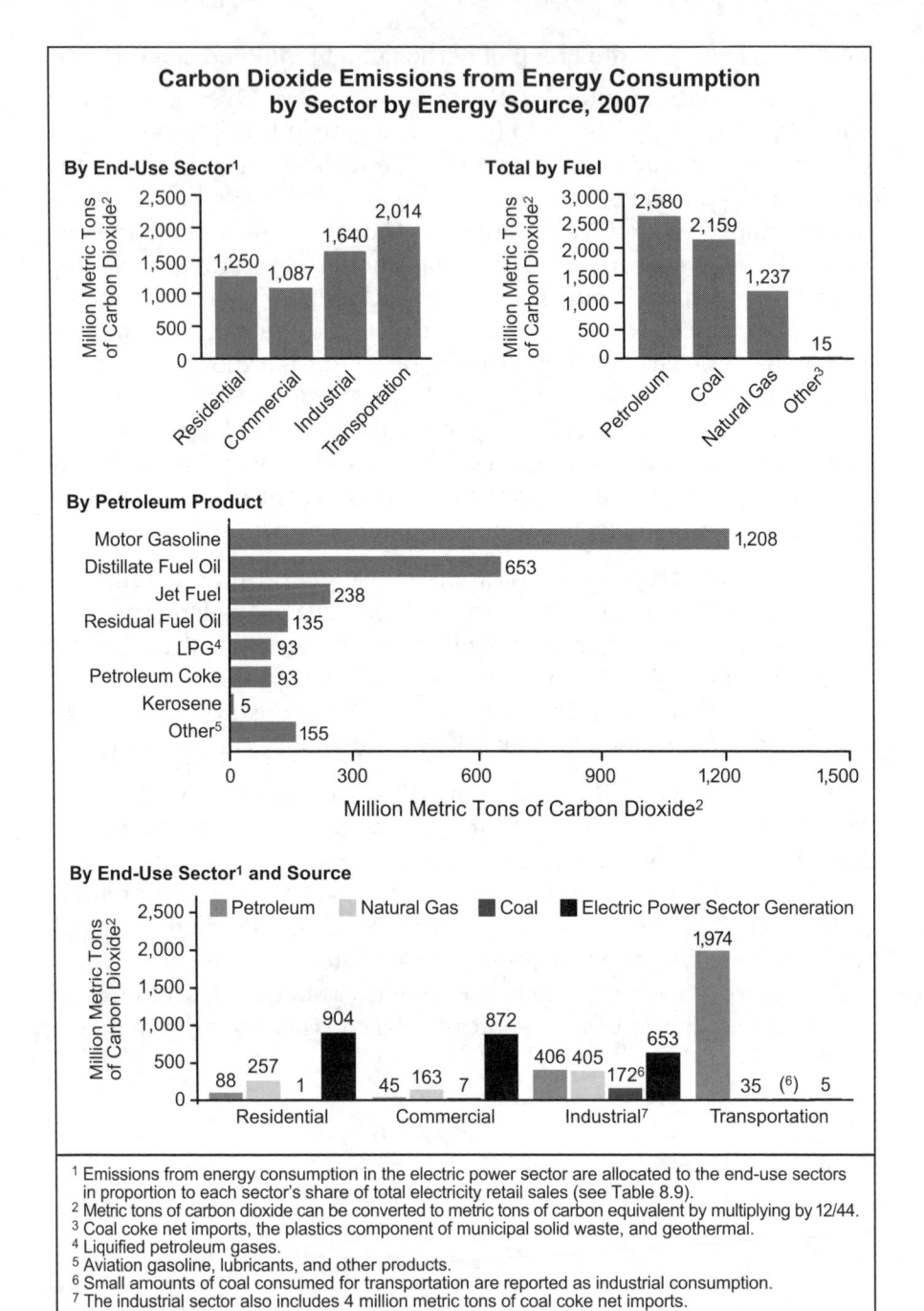

Carbon Dioxide Emissions from Energy Consumption by Sector by Energy Source, 2007

By End-Use Sector[1]
(Million Metric Tons of Carbon Dioxide[2])
- Residential: 1,250
- Commercial: 1,087
- Industrial: 1,640
- Transportation: 2,014

Total by Fuel
(Million Metric Tons of Carbon Dioxide[2])
- Petroleum: 2,580
- Coal: 2,159
- Natural Gas: 1,237
- Other[3]: 15

By Petroleum Product
(Million Metric Tons of Carbon Dioxide[2])
- Motor Gasoline: 1,208
- Distillate Fuel Oil: 653
- Jet Fuel: 238
- Residual Fuel Oil: 135
- LPG[4]: 93
- Petroleum Coke: 93
- Kerosene: 5
- Other[5]: 155

By End-Use Sector[1] and Source
(Million Metric Tons of Carbon Dioxide[2]) — Petroleum, Natural Gas, Coal, Electric Power Sector Generation

- Residential: 88, 257, 1, 904
- Commercial: 45, 163, 7, 872
- Industrial[7]: 406, 405, 172[6], 653
- Transportation: 1,974, 35, (6), 5

[1] Emissions from energy consumption in the electric power sector are allocated to the end-use sectors in proportion to each sector's share of total electricity retail sales (see Table 8.9).
[2] Metric tons of carbon dioxide can be converted to metric tons of carbon equivalent by multiplying by 12/44.
[3] Coal coke net imports, the plastics component of municipal solid waste, and geothermal.
[4] Liquified petroleum gases.
[5] Aviation gasoline, lubricants, and other products.
[6] Small amounts of coal consumed for transportation are reported as industrial consumption.
[7] The industrial sector also includes 4 million metric tons of coal coke net imports.

Fig. 10–2. Carbon dioxide emissions by fuel
(Source: http://www.eia.doe.gov/emeu/aer/pdf/pages/sec12_6.pdf)

Over the past decade during the Kyoto Protocol's existence, there has been a significant increase of CO_2 emissions from activities in China. However, under Kyoto there has been no responsibility for carbon reduction assigned to developing countries. In June 2008, just before the G8 conference, India rejected any commitment to mandatory GHG cuts. There was no movement at the December 2009 Copenhagen conference, and in January 2010, China and India announced some planned future reductions in their carbon intensity per unit of production, which included substantial planned increases in carbon emissions.

The non-Annex I developing nations can host CDM projects sponsored by industries or speculators in carbon from outside their borders. The more than 40% of world GHG emissions emitted just by the United States and China, neither of which country has been covered by limits under the Kyoto Protocol, alone dwarf the cumulative 28% of world GHG emissions attributed to all 38 regulated Annex I Kyoto parties together. Just these two uncovered countries amply exceed by 50% the emissions of all covered carbon emitting countries. Without inclusive active involvement of more than the existing 38 Annex I Kyoto countries, there is no way to stabilize, let alone decrease, atmospheric carbon concentrations under any mathematical calculation, unless one invents a new math which defies physical reality.

Global CO_2 emissions are rising at the rate of approximately 10% per year.[4] More than one-third of CO_2 emissions are attributable to the electric power sector (see chapter 3). Energy use, and the construction of fossil fuel-fired power generation facilities, is increasing as population growth and development continue, especially in developing nations (see chapter 4).[5] The majority of energy and power generation expansion will occur just in Asia over the next decades (see chapter 3).[6] Unabated, this exponential increase in power demand in developing nations will tip the global environment thermostat to runaway global warming risk, regardless of what the United States and other developed nations do to reign in their carbon emissions. If not addressed, the annual increase in GHG emissions in India, China, Brazil, Indonesia, or other of several dozen fast-growing nations, will cumulatively swamp all of the collective GHG reductions of the developed nations complying with the goals of the Kyoto Protocol.

The balance is changing rapidly, in ways that disadvantage the effectiveness of the existing Kyoto Protocol. World CO_2 emissions, currently estimated at about 25 gigatons (Gt) annually, are slightly dominated by OECD nations compared to emissions of developing countries. The crossover point is expected by 2020, and by 2030, projections are that the position of developed and developing nations will have reversed, with developing countries providing the dominant share of CO_2 emissions, and increasing over time into the foreseeable future.

The United States Energy Information Administration (EIA) forecasts a 50% worldwide increase of carbon emissions between 2005 and 2030 as the most likely reference scenario.[7] The International Energy Agency forecasts a 25–90% increase over the same period, concluding that absent a major policy change, CO_2 emissions could increase 130% by 2050.[8] Therefore, these forecasts have in common a scenario that will push the world well past the climate tipping point. A much larger denominator of participating and regulated countries is imperative.

The roles of currently exempt Kyoto countries and urbanization. The trend of world development is important in assessing the possibility of the Kyoto Protocol as an effective or ineffective long-term international climate control architecture. Developing nations are expected to emit a majority of CO_2 emissions within about 10 years, while China has now surpassed the United States as the largest CO_2 emitter in the world. China also had the highest emissions in the world per unit of gross national product (GNP) by a factor more than double other major nations. According to an article in *China Daily*, "In 2005, China's energy consumption per unit of GDP was just more than three times the level of the United States, more than five times that of Germany and eight times that of Japan."[9] Most of the projected increases will occur in developing countries, projected to grow five times as fast as those from industrialized countries over the next 25 years.[10]

By 2030, developing countries are forecast to exceed CO_2 emissions from developed countries. Until 2030, worldwide use of coal is expected to increase roughly 60%. The critical reality is that developing nations represent the geographic core of expected exponential increase in CO_2 emissions, and are where new global warming architecture and law must focus. To bypass or ignore them, as the Kyoto Protocol entirely does, is to invite broad policy failure.

As discussed in chapter 4, much of this shift in balance is dictated by changing world population. World population could reach 8 billion people by 2020 and 9–10 billion by 2050.[11] East Asia (including China) and South Asia now contain more than 3 billion of the world's 6 billion population. At the beginning of the 20th century, there were very few world cities with more than 1 million persons; at the beginning of the 21st century, there were 400 world cities of 1 million or more persons. By 2025, one-quarter of the world's population will be living in Asian cities.

Asia becomes a focus, but the carbon limitations of the Kyoto Protocol only regulate Japan among all Asian countries.

There is a direct relationship between ongoing population shifts augmenting the degree of urbanization and increased carbon emissions. A recent study indicates that while the impact of population growth on emissions is above unity and only slightly different for upper-, middle-, and low-income countries, additional demographic variables (namely, urbanization) demonstrate a very different impact on emissions for low- and lower-middle-income countries with elasticities of emissions and urbanization above unity. This contrasts with upper-middle-income countries with elasticities of emissions and urbanization of 0.72.[12] In upper- and upper-middle-income countries, this elasticity actually is negative. Thus, urbanization in lower-income countries has a greater increase in environmental emission impact in GHG emissions than in higher-income developing countries.

This heterogeneous global warming impact from urbanization in different developing countries by income has implications for carbon policy and technology transfer to mitigate carbon concentrations. Two-thirds of all additional future energy demand could emanate from just China and India.[13] Projections estimate that by 2030, China's GHG emissions will quadruple and Asia alone will emit 60% of the world's carbon emissions.[14]

The world may have to achieve a reduction of CO_2 of 50% or more during the 21st century. Therefore, the new international architecture to control warming must draw a regulatory legal line around major developing nations. The Kyoto Protocol does not do this, and talks of change to Kyoto's line of inclusion have met resistance from developing nations.

Reform 2: Shifting power technology choices in Annex I developed countries

Now it's time to discuss the second necessary reform of the Kyoto Protocol. Only Annex I developed nations are covered by Kyoto regulation. However, CDM offset credits were created as a means to engage resistant developing countries and to facilitate more cost-effective credits for Annex I countries subject to Kyoto caps (see chapter 5). The inclusion of offsets in a cap-and-trade system offers several advantages:

- It allows lower-cost reduction opportunities outside the capped countries to be pursued as lower-cost reduction options.

- Economic sectors that are covered by the carbon emissions caps can be the source for reductions. This can include emission sources not otherwise cost-effectively addressed.
- Offsets can be credited for early reductions or for innovative technologies.
- Offsets can promote technology transfer to developing countries.

The Kyoto CDM architecture has facilitated two significant program design problems. First, the CDM process (see chapter 5) is maximizing creation of imported credits, while not affecting the necessary structural change in the carbon-intensity of the energy production base of either developed or developing nations. Second, there has been a significant overestimation problem in the valuing of imported CDM carbon mitigation credits into developed nations.

Shortcomings of the current architecture. There is an obvious connection between renewable power options and carbon reduction strategies. At current rates of power development, energy-related CO_2 emissions in 2050 would be 250% of their current levels under the business-as-usual pattern.[15] Unprecedented deployment and substitution of renewable energy generation alternatives will be required to alter this trend.

The technology exists to accomplish this (see chapter 3). Converting about 1–2% of the appropriate land area of the Earth to utilize solar energy could satisfy much of the Earth's electricity requirements when solar radiation is available. Annex I countries were anticipated to shift a significant percentage of their power generation bases to renewable energy. It was also expected that the Kyoto CDM mechanism would result in promoting renewable power in developing nations. Neither has been the case. Instead, in developing countries, CDM architecture is trapping methane and other gases and flaring it, without turning it into low-cost electricity in the process.[16] These easy solutions reduce GHGs, but perpetuate the need for electricity for the community from other sources. Therefore, while the Kyoto Protocol CDM process encourages carbon reduction in developing countries, it has not resulted in the substitution of renewable power for conventional fossil fuel power.

Most CDM project creation of certified emission reduction (CER) offsets, which must be sited in (non-Annex I) developing countries, is accomplished by projects to reduce HFC-23, a refrigerant. CERs to date address high GWP industrial gases such as trifluoromethane (HFC-23) and N_2O, as well as CH_4 emitted by landfills and concentrated animal feeding operations (CAFOs). Two relatively obscure industries—adipic acid and chlorodifluoromethane (HCFC-22) production CDM projects dominate. Adipic acid is the feedstock for the production of nylon-66 and releases abundant N_2O as a production byproduct. HCFC-22 has two major applications. It is one of two major refrigerants that were phased in to replace the chlorofluorocarbons (CFCs) under the Montreal Protocol on Substances that Deplete the Ozone Layer. HCFC-22 is also the primary feedstock in the production of DuPont Teflon. These two relatively small industries represent nearly 55% of the supply of issued CERs in the CDM to date. Indeed, the industrial gas emissions that account for one third of CDM reductions do not even occur in the developed world, not because of an absence of adipic acid or HCFC-22 manufacture, but because industries abated them voluntarily and destroy them.

All HFCs collectively constitute less than 1% of GHGs, but they have received almost half of the investment in Kyoto CDM offset mitigation dollars.[17] CDM projects to date have located in a limited number of countries and only a few gases, with "little contribution to sustainable development."[18] By January 1, 2008, more than 1,150 million tons (Mt) CO_2 equivalent (CO_2e) had been registered for delivery through the CDM by the end of the first compliance period in 2012.[19]

A small number of very large projects dominate the supply of CERs from registered projects. In fact, the 45 largest projects (5% of the total number) represent 64% of the total supply to the end of the first commitment period. One country, China, generated more CERs than all other countries cumulatively in the world.

The current Kyoto Protocol architecture does not require the installation of renewable technologies anywhere, but rather only requires the reduction of carbon emissions, which may or may not involve the installation of renewable generation. The impact of CDM projects has not been to promote appropriate renewable investments in developing countries, but rather has served to create additional credits for traditional regulated emitters in Annex I countries. A report by the World Wildlife Fund found that many CDM programs fail to support sustainable development in host CDM countries. Other countries' regulatory techniques to promote renewable power through differential wholesale pricing of power resources, through so-called feed-in tariffs, are not adaptable to the U.S. legal system as experienced by several U.S. states now attempting this (see chapter 17).

Preference for CDM offset trades rather than GHG reduction in EU Annex I countries. Have significant GHG annual emission reductions been facilitated in Annex I countries by the Kyoto architecture? EU carbon emissions are continuing to rise, in part due to the EU carbon trading program (see chapter 6). In 2006, U.S.-monitored carbon emissions fell 1.3% and EU emissions fell 0.3%. Emissions of carbon dioxide in the European Union actually rose 1.1% in 2007 compared to 2006.[20]

The data on most leading EU countries does not show the trend line that one might have anticipated since the 1992 Rio Declaration commencement of the assault on GHGs emissions. As of a recent assessment contrasting 2006 CO_2 emissions with 1990 baseline emissions, Austria, France, Ireland, Italy, and Spain were double or more over time, while Finland, Germany, the United Kingdom, and Sweden were higher.[21] As a whole, EU countries were up about 25%. CO_2 emissions have increased particularly in cement production, petroleum refining, electricity and heat production, and transportation.

According to an official with the World Wildlife Fund, the European Union is playing "tricks on the atmosphere" in claiming that it is making significant carbon reductions: "the European Union in pretending to take, compared to what it means in reality, 20% emissions reduction by the EU by the year 2020 on 1990 levels means no more than 4-5% emissions reductions domestically," from 2009 onward.[22] Of this 20%, 8% is already accounted for by the deindustrialization by former Soviet states, and the bulk of the remainder is achieved by foreign CDM projects rather than in Annex I countries.

In Europe, there is continued use of low-cost coal, rather than a move to renewable power resources. During Phase I of the EU-ETS (2005-2007), former Eastern block countries received a carbon allocation surplus of free allowances based on a 1990 baseline pre-Soviet restructuring, which allowed their industries to sell surplus free allowances for profits. Romania and Bulgaria demanded larger CO_2 emission allocations than they were given by the European Commission, and thereafter judicially appealed this allocation to the European Court of First Instance.[23] Poland, the Czech Republic, Hungary, Romania, and Slovakia launched legal actions against the EU Commission, asserting that even their relatively modest future carbon limits are too strict for the countries' economic growth.[24]

Prime Minister Ivars Godmanis of Latvia stated that he would veto any climate package unless it contained more concessions for the Eastern countries that joined the EU in 2004.[25] "Poland is ready to veto," stated Poland Foreign Minister Radek Sikorksi.[26] Post 2012, the EU proposal, reflecting accession to political pressure to achieve consensus, now is to allow central and eastern EU countries to increase emissions up to 20% above 2005 levels, rather than reduce carbon.[27] Political concessions were made in Poznan, Poland, meetings in December 2008, where the former Eastern-block EU states were given 12% of future allowance auction revenues to subsidize existing high-carbon industries.[28] A three-year delay was given to the auto industry, and other select industries in key countries.

Non-EU Annex I Kyoto Protocol countries have not been particularly successful in reducing GHGs. Canada was sued by environmental groups for failure to honor Kyoto commitments, making it the first country to be sued under the Kyoto Protocol.[29] There have been myriad GHG lawsuits in the United States, but not pursuant to the Protocol, which the United States has not yet ratified (see chapter 11).

Three of the five largest GHG-emitting countries are not covered by Kyoto limitations. Russia, among the two covered in those top five, is covered but has extremely generous limits that will require no action to reduce carbon emissions from Russia's 1990 Kyoto point of reference. Kyoto has not facilitated any significant carbon reduction internally in the bulk of its Annex I countries. Kyoto has allowed increases in Annex I CO_2 emissions through the CDM to import additional offsets, which add to allowed emissions.

The existing CDM projects will generate an estimated 2.6 billion CERs by 2012.[30] This would equate the value of offsets to almost 10% of monitored emissions in regulated Annex I countries, which is a huge tradable quotient. Each CDM CER generated in a developing country increases the GHG emissions allowed in an Annex I country to which it is exported through the Kyoto registration process. It is difficult to conclude that the EU-ETS carbon system that parallels the Kyoto Protocol resulted in any carbon reductions that would not have occurred in the absence of the cap-and-trade system, according to Rachel Miller, director of federal affairs for British Petroleum Corporation.[31]

Reform 3: Creating international enforcement mechanisms

Lack of enforceable controls. Enforceability is a key element of any regulatory regime, and it is the third area of necessary reform highlighted here. A significant problem with Kyoto, as with some other international mechanisms, is that it is a wholly voluntary agreement.[32] There is no provision in the Kyoto Protocol to ensure compliance of any nation that fails to achieve its reductions or violates any provision. There is no effective international organization with any effective power over carbon-emitting nations. Achievement of goals, at the end of the day, is voluntary and unenforceable. Given the timing pressure to achieve GHG reductions, there needs to be a significant fix (see chapter 4).

The EU-ETS system is different in this important regard from the Kyoto Protocol. In the EU-ETS, as noncompliance occurs, the EU Commission can refer noncompliance to the European Court of Justice, and there is more centralized enforcement power. The EU covers 27 of the 38 Kyoto-regulated countries with a parallel regulation, but the EU only covers CO_2 emissions, not the other five groups of GHG chemicals (see chapter 2). With many countries slated to miss their Kyoto goals, there is no sanction, no penalty, and no accountability in the international Kyoto architecture. Unlike some international legal disputes, there is not even an international court to which to take Kyoto disputes.

Trade sanctions. Comparative competitiveness concerns are a major worry of nations when revising the Kyoto architecture. Comparability in undertakings and commitment is an important requirement for the inclusion in the post-Kyoto regime. The trade sanction has emerged as a discussed surrogate weapon of choice in countries contemplating sanctions against other nations that do not abate GHG emissions. The 2008 U.S. Lieberman-Warner carbon regulation bill, which was not enacted, would have required importers to obtain carbon allowances to cover goods imported from countries without carbon restrictions, starting in 2014. This would substantially increase the cost of importing such goods. The 2009 Waxman-Markey carbon bill in the United States opens up the possibility of tariffs on imported high-carbon goods from countries with no comparable carbon controls.

Trade sanctions, sometimes known as *border adjustment measures* are likely to become part of future carbon restrictions in a number of key nations. These sanctions are designed to push developing countries that have not participated, avoid free rider incentives for developing countries, and prevent carbon leakage from unregulated countries to regulated ones. These sanctions would primarily cover manufactured items. However, nations may not be discouraged from converting trade sanctions into the surrogate weapon of climate policy. Governments of industrial nations have export-import banks and organizations such as Overseas Private Investment Corporation (OPIC) that could, positively, emphasize subsidy or support of low carbon technologies in lieu of sanctions.

Current Energy Secretary Steven Chu also supported tariff duties imposed by the United States on carbon-responsible imports from a non-regulated country.[33] The flip side of this proposition is where the debate in the United States has veered: giving away carbon emission allowances to U.S. industry that is particularly vulnerable to foreign competition. China's top climate negotiator, Li Gao, disagreed, stating that since China manufactures and exports many projects, "this share of emissions should be taken by the consumers, but not the producers."[34]

Such sanctions could violate provisions of the General Agreement on Tariffs and Trade (GATT), which established the World Trade Organization.[35] GATT is the first multilateral, legally enforceable agreement dealing with trade and investment in services. Quantitative limits on new imports are prohibited, internal taxes on foreign products in excess of those on domestic products are not permitted, and affording most favored nation status to imports is required.[36] Exceptions are provided for measures necessary to human life or health, or relating to conservation of exhaustible natural resources that also affect domestic production.[37] These general prohibitions may apply to the atmosphere as an exhaustible natural resource if they are applied in a manner of arbitrary discrimination.

The failure to impose penalties under Kyoto stands in contrast to EU-ETS carbon scheme and U.S. penalty provisions for other emissions than carbon. Because the U.S. cap-and-trade criteria pollution system has an automatic penalty for non-compliance that is significantly higher than the market price for acquiring allowances for the pollutants SO_2 and NO_x, coupled with a liquid market for trading allowances, there is very high compliance with required emissions levels with the U.S. SO_2 and NO_x trading programs. In 10 years of operation, there have been only 21 excess emission penalties, plus nine additional civil penalties for other violations such as failure to monitor and report emissions. These mechanisms have worked. The EU-ETS scheme provides compliance penalties of €100 per ton between January 2008 and December 2012 for failure to have enough AAUs.

The Kyoto Protocol. However, in contrast, violation of the Kyoto carbon pollutant targets carries no penalty. This recourse to enforcement tools is important because otherwise there will be slippage between the carbon goals and the achievements. Even with penalties, European GHG emissions in industrialized European countries are increasing. Many EU countries could miss their Kyoto targets. Canada and Japan are projected to miss their interim 2010 targets by 500 million tons each of CO_2; Japan's emissions are rising, and Canada has backed away from its target obligations.[38] There is no international enforcement mechanism.

The issue of lack of enforcement in Kyoto and effective pressure regarding EU-ETS sanctions has even begun to erode consensus among those countries that have joined to be regulated by the Kyoto Protocol. The advantages of being a trading member of the unified European Union have caused many former Eastern block countries to seek membership, and then to chafe under the pressures for, and threaten to boycott, carbon reduction (see chapter 6). Poland and Bulgaria, both recently admitted to the European Union and former Eastern block countries, argue that more advanced Western EU countries should do more carbon reduction, while the poorer Eastern European countries should do less. Even Italian Prime Minister Silvio Berlusconi stated that the proposed EU-ETS carbon reduction targets for 2020 would "crucify Italy."[39]

Poland generates 97% of its power production from coal, and in December 2008, EU nations tentatively agreed to extend the allocation of free carbon emission allowances to utilities equal to the needs of an efficient coal-burning power plant and double time by extending compliance dates to 2020. This threatens even the modest goals of the Kyoto Protocol and the EU-ETS. Yet, some of the Eastern European countries (Poland, Czechoslovakia, Hungary) are expected to continue challenge of their future allocations. Central and eastern EU states have launched legal proceedings against the European Commission, alleging their allocations are already too low. In the Kyoto Protocol, there is no enforcement provision.

The United States, which is not an Annex I participating party of the Kyoto Protocol, would miss its Kyoto limit (if it applied) by 1,500 million tons of CO_2.[40] There is little other prospect of the United States changing this vector of increasing carbon emissions or closing this huge deficiency before the 2012 end of the Kyoto compliance period, regardless of any legislative efforts at the federal or state levels. Among the U.S. federal legislative proposals debated, none would have had any substantial bite or impact whatsoever by 2012. The next chapter looks more closely at this lack of legislative or executive bite at the federal level by examining the role of the courts in setting carbon emission obligations.

Notes

1. Harris, Paul G. 1999. Common but Differentiated Responsibility: The Kyoto Protocol and United States Policy. *Environmental Law Journal.* New York University. 7: 27, 34.

2. UNFCCC Kyoto Protocol. Articles 3(1), 4(1) ("All Parties, taking into account their common but differentiated responsibilities..."). The concept was originally part of the Montreal Protocol on Substances that Deplete the Ozone Layer, Sept. 16, 1987, 1522 U.N.T.S. at Article 5 and at Annex II, paragraph T, Article 10 (as amended in the London Amendments, June 29, 1990, UNEP/OzL.Pro2/3).

3. Blair, Tony. 2008. *Breaking the Climate Deadlock: A Global Deal for our Low-Carbon Future.* The Climate Group. 6.

4. See, Purdy, Ray. 2006. The Legal Implications of Carbon Capture and Storage under the Sea. *Sustainable Development Law & Policy.* American University, Washington College of Law. (Fall): 22.

5. World Bank. Statement, Ministerial Segment – COP11 – Montreal 4. http://siteresources.worldbank.org/ESSDNETWORK/Resources/MINISTERIALSEGMENTCOP11Montreal.pdf; International Energy Agency. 2004. *World Energy Outlook 2004.* www.worldenergyoutlook.org.

6. International Energy Agency. 2004. *World Energy Outlook 2004.* www.worldenergyoutlook.org.

7. U.S. Department of Energy. 2008. *EIA International Energy Outlook 2008.* www.eia.doe.gov/oiaf/ieo/index.html.

8. UN International Panel on Climate Change. 2007. *Fourth Assessment Report.*

9. China Daily. 2007. Energy Consumption Per Unit of GDP Continues to Fall. *China Daily.* July 15.

10. U.S. DoE, *EIA International Energy Outlook 2008.*

11. United Nations. Department of Economic and Social Affairs, Population Division. 2006. *World Population Prospects: The 2006 Revision.* 5. http://www.un.org/esa/population/publications/wpp2006/English.pdf.

12. Fondazione Eni Enrico Mattei (FEEM). 2008. *The Impact of Urbanization on CO_2 Emissions: Evidence from Developing Countries.* Social Science Research Network Electronic Paper Collection. http://ssrn.com/abstract=1151928.

13. Ibid.

14. Cooper, Deborah. 1999. The Kyoto Protocol and China: Global Warming's Sleeping Giant. *Georgetown International Environmental Law Review.* 11: 401, 405.

15. International Energy Agency. 2006. *Energy Technology Perspectives: Scenarios and Strategies to 2050.*

16. CDM projects are credited by capturing or destroying carbon gases, rather than maximizing efficiency. Because one receives greater credit by 2,100% from destroying methane than CO_2, many CDM projects capture rural or agricultural methane and flare it (converting it to CO_2). But in that flaring process, it is not used

to produce electricity, which is otherwise locally supplied by traditional sources. Rather than make the lifecycle cost-effective investment in electric generation technology, CDM investors often minimize capital investments by flaring methane and ignoring essentially free-at-the-margin electric generation. The source of this is observation of the author in his extensive work around the world advising international organizations and private entities on carbon policy.

17. Ball, Jeffrey. 2007. Kyoto's Caps on Emissions Hit Snag in Marketplace. *Wall Street Journal*. December 3, A-1.

18. El-Ashry, Mohammed T. 2008. Framework for a Post-Kyoto Climate Change Agreement. *Sustainable Development Law & Policy*. American University, Washington College of Law. (Winter): 2, 5.

19. U.S. Government Accountability Office. 2008. *International Climate Change Programs: Lessons Learned from the European Union's Emissions Trading Scheme and the Kyoto Protocol's Clean Development Mechanism*. GAO-09-151: 35, figure 4.

20. Bureau of National Affairs. 2008. Analysis Shows Increase in EU Carbon Emissions. *Environment Reporter*. April 11, 706; see also http://ec.europa.eu/environment/ets/.

21. UNFCCC Secretariat. 2008. *Annual European Community Greenhouse Gas Inventory 1990–2006 and Inventory Report 2008*. Technical Report No. 6/2008, 175, table 3.61.

22. EurActiv. 2009. EU 'cheating' the world on climate, says WWF. *EurActiv.com*. April 14. http://www.euractiv.com.

23. Convery, F., D. Ellerman, and C. de Perthuis. The European Carbon Market in Action: Lessons from the First Trading Period. *Global Change Science Policy Report No. 162*. Massachusetts Institute of Technology. (June): 22.

24. Herrero-Martinez, Erica. 2007. States Study Carbon Trading. *Wall Street Journal*. August 1: B5A.

25. Castle, Stephen. 2008. European Nations, Fearing Downturn, Seek to Revise Agreement on Emissions Cut. *New York Times*. October 17: A6.

26. Phillips, Leigh. 2008. Italy, Poland Threaten to Veto EU Climate Package. *EUobserver.com*. October 16.

27. Goldirova, Renata. 2008. Brussels to Unveil EU Green Strategy Amid Strong Criticism. *EUobserver.com*. January 21.

28. Harrison, Pete and Huw Jones. 2008. EU Finalizes Deal to Fight Climate Change. *Reuters*. December 16.

29. Song, Vivian. 2008. Canada Charged with Kyoto Noncompliance. *Sarnia, Ontario Observer*. June 19.

30. Voigt, Christine. 2008. Is the Clean Development Mechanism *Sustainable? Sustainable Development Law & Policy*. American University, Washington College of Law. (Winter): 15.

31. Carbon Control News. 2008. Cracking Down on Offset Projects. www.carboncontrolnews.com. August 6.

32 IPCC. 2001. *Third Assessment Report: Climate Change 2001.* http://www.ipcc.ch/pub/online.htm.

33 Environment and Energy Daily. 2009. Carbon Tariffs an Option for Obama Admin. http://www.eenews.net. March 18.

34 Associated Press. 2009. China Hopes Climate Deal Omits Exports. *Associated Press.* March 17.

35 United Nations. 1994. Marrakech Agreement Establishing the World Trade Organization, Annex 1A, 1867 United Nations Treaties Series 187, 33 International Legal Materials 1153.

36 General Agreement on tariffs and Trades (GATT), art. 3: 2.

37 GATT, art. 20(g).

38 Ball, *Kyoto's Caps on Emissions Hit Snag in Marketplace*, 19.

39 Charter, David and Rory Watson. 2008. Black Clouds Hang over Green Targets as EU States Say We Can't Afford Them. *Times.* (U.K.) October 17: 6.

40 See IPCC Web site at www.ipcc.ch; Purdy, *The Legal Implications of Carbon Capture and Storage under the Sea*, 22.

11) THE FULCRUM LEVERAGE ON GLOBAL WARMING: ROLE OF THE COURTS

The U.S. Supreme Court Climate Ruling

The rule of law weaves through all climate change issues. Regulation of behavior and conduct of carbon emitters is a legal matter. Similarly, litigation and liability risk now include carbon issues. This risk affects various businesses and government actions. Climate-related obligations are being interpreted by the courts in response to suits by environmental groups, citizens, and state governments. These risks from litigation are wide-ranging and accelerating.

Both state courts and federal courts, as well as international tribunals, have adjudicated climate related disputes. In 2007, a major decision was issued by the U.S. Supreme Court, but there was extensive litigation in other matters and other places even before the Supreme Court acted (see next section). In the Supreme Court case of *Massachusetts v. EPA*, a suit was filed by 12 states and several cities against the U.S. Environmental Protection Agency (EPA) to force it to regulate carbon dioxide and GHGs associated with vehicle emissions.[1]

The Supreme Court held that the EPA has authority under the Clean Air Act (CAA) to regulate GHG emissions from new motor vehicles, and that even indirect harm from climate change can confer standing on these plaintiffs to sue the EPA for failure to comply with CAA requirements. Note that the Court was not asked to hold that GHG emissions from stationary sources like power plants are regulated by the CAA, that auto companies are liable for the harm caused by GHG emissions, or that the EPA is required to regulate GHG emissions from mobile sources. It waited until the Obama administration for the EPA to decide to actually regulate CO_2.

A lasting important impact of this decision is to lower the standard that government litigation plaintiffs must show to demonstrate causation and redressability and to demonstrate the necessary environmental standing to enable them to initiate an action before the courts.[2] The Court found that although the effects of global warming are "widely shared," Massachusetts had legal standing because Massachusetts could show some modicum of harm, thereby rejecting the EPA's argument that global warming's widespread effects negated standing for any particular individual plaintiff. Massachusetts owns, operates, and maintains 53 coastal parks, numerous coastal recreational facilities with significant infrastructure combined with roads, walkways, sea walls, pump stations, and piers where the state alleged that damages from global warming could run into the hundreds of millions of dollars. The Court found that because Massachusetts owns or has interests in significant coastal land that allegedly will be affected by sea level rise from global warming, it could show the requisite particularized harm to maintain suit against a federal agency.

The reduction of new vehicle emissions was significant enough to affect global warming, given that the transportation sector alone contributes one-third of the greenhouse gases emitted in the United States, and along with power plant emissions, dominates GHG emissions worldwide. The Court held that just because other countries such as China and India are poised to increase greenhouse gas emissions does not mean a reduction in the United States would have no effect. The Court explained that while regulating motor vehicle emissions may not, itself, reverse global warming, redressability requirements linking the requested relief with abatement of the alleged harm were met where regulation would impact, even to a small degree, global warming gases. Especially with global pollutants, such as CO_2, this causal link is difficult to demonstrate because of the indirect relationship between CO_2 emissions and direct local impacts. The Court noted that while the risk of catastrophic harm is remote, it is nevertheless real and could be reduced if relief were afforded by the Court.

In a subsequent decision to the 2007 Supreme Court decision in *Massachusetts v. EPA*, the United States Court of Appeals for the District of Columbia Circuit, which has primary jurisdiction over challenges to government regulations, in 2009 held that the 2007 Supreme Court decision on standing applies only to state sovereign interests, while the standing of citizens or environmental groups to sue on climate change is not as expansive or permissive.[3] Notwithstanding, this later decision preserved rights for an environmental group to sue regarding government agency violation of *procedural* obligations that an agency did not properly follow. However, the court of appeals held that environmental group plaintiffs lacked standing to sue on the *substantive* merits of their claims because they could not prove that they suffered more actual and imminent harm than others due to climate change.

The court of appeals in this 2009 opinion also found that there was no showing by plaintiffs, as required, that government action was the proximate cause of the global warming. Instead, the court characterized the plaintiffs as speculating about the attenuated linkage alleging consumers would use additionally produced oil resulting from more leasing of drilling rights to cause release of carbon in the atmosphere in the future. The decision returns citizen plaintiffs to the earlier standard for environmental group standing articulated by the Supreme Court in *Lujan v. Defenders of Wildlife*, 504 U.S. at 564.

Of note, actions of the defendant U.S. Department of the Interior outer continental shelf leasing activities have a less direct effect on climate change than actions of other agencies that more directly affect the burning of fossil fuels in power production, heating, or transportation. The 2009 decision also raised again the lack of *ripeness* of the claim. Ripeness is a discretionary court doctrine that requires plaintiffs to wait until there is final agency action that creates harm to ask for court intervention. If plaintiffs litigate too early, the claim could not yet be ripe, while waiting too long could impede timely action and moot the legal claim.

Early Court Challenge on Climate Change

The status of climate change litigation in the United States is such that no stakeholder was immune even *before* the 2007 Supreme Court decision. These challenges were brought by environmental groups, state governments, and citizens, and raised a variety of legal issues. It is interesting to note who were the plaintiff challengers, and who were the target defendants. These suits often raise common law legal complaints. The substance of the climate-related damages alleged also is of note. This section will survey some of the more noteworthy carbon litigation.

Federal and state government regulators were the targets of suit by an individual in *Korsinsky v. EPA*, a 2005 matter in the U.S. District Court for the Southern District of New York.[4] A complaint was filed by a New York resident who claimed that global warming is a public nuisance under federal common law and New York statutory and/or common law.[5] In *Korsinsky*, a *pro se* plaintiff alleged that global warming would physically injure him over time, causing him to suffer sinus-related diseases enhanced by the risk of contaminated drinking water caused from increased floods. The plaintiff claimed that he "developed a mental sickness" because he was so worried about what might happen to him because of global warming. As relief, plaintiff requested that defendants be held jointly and severally liable and also that the court require defendants to implement his invention.

The court dismissed plaintiff's claims because he lacked standing. Plaintiff's physical injury did not rise to the level of a *"certainly impending"* injury required for the U.S. Constitution's Article III case or controversy standing, and his alleged mental injury was not specific enough, nor would implementing plaintiff's invention redress the harm of global warming. The court of appeals affirmed, holding that plaintiff had not established standing because he had not explained exactly what possible injury had been caused by the appellants' actions nor did he show how the injury could be redressed, therefore the court could not grant jurisdiction.[6]

Federal permitting agencies also were the subject of litigation in *Coke Oven Environmental Task Force v. EPA*[7] and in *New York v. U.S. EPA*,[8] cases in the United States Court of Appeals for the District of Columbia Circuit. In *Coke Oven*, at issue was a petition for review challenging EPA's decision not to regulate CO_2 emissions for the purposes of global climate change. Three environmental groups, 10 states, and two cities originally petitioned the EPA to enact standards regulating greenhouse gas emissions for new stationary (power plant and large industrial) sources. The case was stayed pending the Supreme Court's decision in *Massachusetts v. EPA*. After that decision in 2008, the parties signed a settlement agreement under which the EPA agreed to either issue a notice of new rulemaking or direct final rulemaking imposing restrictions on emissions from large coal- or oil-burning steam power plants. The new rules must require new or newly modified large power plants to drastically limit their emissions.

State or local government permitting agencies were targeted by environmental groups and the state in *Center for Biological Diversity, the Sierra Club, the Attorney General and San Bernardino Valley Audubon Society v. San Bernardino County*.[9] This alleged San Bernardino County violated the California Environmental Quality Act (CEQA), attributable to the county's failure to address the impacts of its long-term land use planning document on climate change, global warming, and GHG emissions. The suit asserted that the county ignored requests from the California attorney general and various conservation groups to assess climate change issues in the development of the plan and the CEQA process. In 2007, the attorney general settled with the county under terms that require the county to develop an inventory of greenhouse gas emissions related to land-use decisions and county operations, set emission reduction goals, and adopt mitigation measures.

Companies that make products that burn fossil fuels were sued by a state in *California v. General Motors Corporation*, in the United States District Court for the Northern District of California.[10] The State of California filed suit against six automobile manufacturers requesting compensation for damage inflicted by their vehicles' GHG emissions. The suit was brought under both the federal and California common law of public nuisance requesting compensation for damages allegedly inflicted by their vehicles' greenhouse gas emissions, as well as a declaratory judgment that the manufacturers will be held liable for any further damages caused by climate change. Although Americans own 30% of the world's cars, because of their efficiency levels and miles driven annually, these cars account for about 45% of the entire global carbon dioxide emissions from vehicles. This suit was dismissed in September 2007 under the political question doctrine, which allows courts to sidestep suits that question decisions committed to other branches of government. The court dismissed without prejudice to the state refiling its nuisance claims in state court. California appealed to the United States Court of Appeals for the Ninth Circuit, which was pending.

Utility companies that emit greenhouse gases were sued by several state attorneys general in *State of Connecticut v. American Electric Power*.[11] This challenged operation of stationary sources, and more importantly, power companies. The State of Connecticut and others filed suit against five utility companies under federal common law and/or statutory common law of the states alleging the tort of public nuisance regarding global warming issues. Plaintiffs alleged that defendants annually emit 650 million tons of carbon dioxide, which amounts to one quarter of the electric power sector's carbon dioxide emissions; America's electric power sector is responsible for 10% of the worldwide man-made carbon dioxide emissions. Plaintiffs requested equitable relief, that defendants be held "jointly and severally liable" for contributing to an ongoing public nuisance of global warming, and be enjoined to cap their emissions of carbon dioxide and then reducing them by a specified percentage each year. The court granted the defendants' motion to dismiss as a non-justiciable political question. This case was dismissed two years prior to the Supreme Court decision in *Massachusetts v. EPA*. The matter was appealed to the United States Court of Appeals for the Second Circuit, which reversed and restored the plaintiffs' standing and gave the suit new judiciable life on its merits.

Similar in both of the two aforementioned decisions, the courts held that the plaintiffs lacked standing because the issue was non-justiciable. In contrasting the decision in *GMC* with *AEP*, the *GMC* court explained that deciding the claims would force the court to balance competing policy interests, which is the type of policy decision to be made by elected political branches, rather than the courts:

> Plaintiff's global warning nuisance tort claim seeks to impose damages on a much larger and unprecedented scale by grounding the claim in pollution originating both within, and well beyond, the borders of the State of California. Unlike the equitable standards available in Plaintiff's cited cases, here the Court is left without a manageable method of discerning the entities that are creating and contributing to the alleged nuisance. In this case, there are multiple worldwide sources of atmospheric warming across myriad industries and multiple countries.[12]

Large industrial CO_2 emitters were the targets of litigation by an environmental group in *Northwest Environmental Defense Center v. Owens Corning*.[13] A complaint was filed alleging that Owens Corning was constructing a manufacturing facility that would emit 250 tons of greenhouse and ozone-depleting emissions without obtaining a required air contaminant discharge permit. In 2006, the court ruled that plaintiffs had standing to pursue their claim, and accordingly denied a motion to dismiss filed by the defendant. On the same date, the parties filed a stipulated order of dismissal, which was approved by the court. Under the stipulation, the defendant agreed to pay $600,000 to fund projects by various nonprofit and education groups, as well as $250,000 in settlement of the plaintiffs' claims for attorneys' fees and costs. The defendant also pledged not to use a certain greenhouse gas in its facility.

Insurers were defendants in the climate change case of *Comer v. Murphy Oil USA* (Nationwide Mutual Insurance Company) (a.k.a. the Hurricane Katrina Litigation).[14] A class action complaint for damages and declaratory relief against several insurance companies was filed by individuals alleging claims based on insurance coverage issues and global warming issues. The claim was based in part on the defendants' emissions, which were alleged to have enabled Hurricane Katrina to develop unprecedented strength and as a result allowed the class to suffer a common set of damages. The court declined to certify the classes and dismissed without prejudice plaintiffs' claims against the insurance and mortgage lending defendant companies, based on standing and non-justiciable political questions. In October 2009, the U.S. Court of Appeals for the Fifth Circuit overturned the decision allowing this public nuisance litigation to proceed on the merits.

Even U.S. credit support institutions have been carbon litigation defendants in *Friends of the Earth v. Mosbacher*.[15] The plaintiff sued alleging that Overseas Private Investment Corporation (OPIC) and Export-Import Bank of the United States (Ex-Im) failed to comply with the National Environmental Policy Act (NEPA) and that global warming is caused by GHG emissions caused from their assistance to fossil fuel projects worldwide. Ex-Im provides financial support for United States exports. Ex-Im and OPIC track and report aggregate GHGs from their respective projects. The plaintiffs identified seven projects funded by OPIC and Ex-Im, asserting that they should have prepared NEPA environmental impact statements. While OPIC's handbook[16] requires a review of whether OPIC credit support would violate any OPIC requirement, this review does not conform to a U.S. NEPA review.[17] In 2007, the court concluded that (1) defendants were not exempt from NEPA, (2) the claims presented did not involve extraterritorial application of NEPA, and (3) material facts remained in dispute such that summary judgment was not appropriate. The court certified various issues in the case for interlocutory appeal to the United States Court of Appeals for the Ninth Circuit.

Finally, in the Inuit petition to the Inter-American Commission on Human Rights, it was claimed by Native American citizens in Alaska that CO_2 emissions constituted a violation of fundamental human rights.[18] This was not in U.S. courts. The Inter-American Commission on Human Rights refused to hear the Inuit petition. Plaintiffs indicated that they will not give up the fight and have asked the commission for further information on why the petition was not going forward.

So even before the 2007 U.S. Supreme Court decision on climate change, every type of entity was at litigation risk, and that risk is now growing. The first wave of litigation on environmental claims of any kind against institutions and companies often is not successful, with prospects evolving over time as the science progresses, documents are discovered, and public opinion changes. The Supreme Court decision in *Massachusetts v. EPA*, while surely not resulting in any immediate judicially imposed carbon restrictions, opened the doors more widely for judicial standing of parties to raise carbon-related claims. While litigation is just beginning, it is possible that GHG emissions could become the next tobacco, asbestos, or MTBE litigation, with the courts finding defendant liability or required actions. In any event, because of the range of litigation, the courts are creating a fulcrum to leverage climate change policy and application.

Recent Legal Claims

Since the Supreme Court 2007 decision regarding CO_2, there has continued litigation and legal action regarding climate change and carbon control in the United States. It is just in the early stages. It is not the purpose of this section to create a scorecard of this litigation status. However, a cross-section of the recent litigation on carbon control following the 2007 U.S. Supreme Court decision in *Massachusetts v. EPA* is highlighted in the following paragraphs to provide a recent snapshot of the context of this legal challenge.

Some litigation picks up where the broad Inuit petition profiled in the previous section left off, with Native American groups bringing direct claims against private sector contributors of GHGs. Claims on global warming can also rebound into secondary insurance claim litigation between the emitters of GHGs and their insurance providers.

In *Kivalina v. ExxonMobil Corp.*,[19] the native village of Kivalina sued 14 electric utilities, 5 large oil companies, and Peabody Coal Company, alleging that defendants' emissions of GHGs collectively and individually constituted both a public and private nuisance and have contributed to climate change. The resulting climate change was alleged to have caused damages to the plaintiffs by resulting in sea level rise and melting of sea ice, destroying Kivalina. Plaintiffs also alleged that the corporate defendants participated in a civil conspiracy to discredit the science of climate change. Plaintiffs claimed that the village of Kivalina must be relocated due to the effects of climate change and that the cost of relocation was estimated to be $90–$400 million. The district court dismissed the nuisance claims in 2009.

This primary litigation related to global warming spawned secondary litigations regarding insurance coverage for both the costs of litigation defense and any underlying damage to property. AES Company, an owner of power plants, was one of the sued electric providers involved in the Kivalina litigation. When initially sued in that primary action, AES tendered the *Kivalina v. Exxon Mobil* climate change lawsuit to Steadfast, its insurance provider under its applicable policy, and Steadfast agreed to defend AES in the primary litigation pursuant to a reservation of rights. In 2008, Steadfast changed its original position on the tendered claim coverage and defense obligation and filed a complaint against its insured for declaratory relief and reimbursement of costs incurred to date.

In *Steadfast Insurance Company v. The AES Corporation*, a declaratory judgment action was brought against AES by its insurance carrier, Steadfast Insurance.[20] The claim was for a declaration that Steadfast did not owe coverage because the Kivalina complaint does not allege property damage caused by an "occurrence" under the policy; in other words, that global warming is not the result of any "accident" that is covered under the policy. The insurance provider plaintiff alleged that given the utility industry's long-standing knowledge of risks associated with greenhouse gases, there was no accident. It also alleged that the emission of greenhouse gases is "air pollution," which is subject to an insurance policy total pollution exclusion that negates policy coverage of any kind for such air pollution. The litigation was still pending at the time of this publication.

In addition, since the 2007 Supreme Court decision allowing EPA to regulate CO_2 from vehicles, the failure of regulatory agencies to address climate change has been the subject of a plethora of suits against government agencies for not adequately considering GHGs. Below are set forth several examples of these suits against various federal agencies and scientific advisory groups to these government agencies. In *California v. EPA*, California sought to compel a decision by the EPA regarding whether California would be granted a waiver under the Clean Air Act to adopt more stringent greenhouse gas emission standards for cars and trucks than are required by the federal government.[21] The parties filed a joint stipulation for dismissal of the action in 2008.

In *East Texas Electric Cooperative v. Rural Utilities Service*, a rural electric cooperative filed a complaint against the U.S. Department of Agriculture, Rural Utilities Service (RUS), as well as the Sierra Club. The complaint alleged that RUS failed to process loan applications requesting approximately $240 million to be used for the purchase of shares in two power plants located in Arkansas.[22] RUS's administration of loan programs under the federal Rural Electrification Act recently had become a target of the Sierra Club and other environmental groups. They alleged that RUS's lending practices violated the National Environmental Policy Act by not adequately evaluating the impacts of financing fossil fueled generation on climate change. This litigation concerning government lending agencies succumbing to environmental group pressure was still pending at the time of this publication.

In another case, but this time from an environmental organization rather than a cooperative utility, RUS was again the target of litigation. In *Center for Biological Diversity v. Rural Utilities Service*, the litigation forum for the suit was moved from California to Kentucky.[23] Several environmental groups sued RUS seeking declaratory and injunctive relief under the National Environmental Policy Act (NEPA), for failure to adequately disclose and analyze the cumulative impacts and environmental risks associated with the financing of transmission line projects when evaluated together with coal-fired electricity generating units. For more on NEPA and environmental impact assessments required in the United States, as well as under various requirements of international bilateral agencies funding power projects, see Steven Ferrey with Dr. Anil Cabraal, *Renewable Power in Developing Countries: Winning the War on Global Warming*, PennWell Publishing, chapter 13.

Scientific advisory groups to the government have been sued involving their failure to address global warming. In *Center for Biological Diversity v. Brennan*, several environmental groups sued to compel federal scientific bodies to prepare periodic scientific assessments of the effects of global climate change and to make research recommendations, as required by the Global Change Research Act.[24] The United States Court of Appeals for the Ninth Circuit granted appellant environmental groups' motion to dismiss the appeal and in 2008 the defendants filed a notice of compliance with the district court's 2007 order requiring their consideration of global climate change in their advisory role.

In *Montana Environmental Information Center v. Johanns*, several environmental groups brought a NEPA challenge to the federal financing of coal-fired power plants.[25] The complaint alleged that the federal government was preparing to lend billions of federal dollars to build several new coal-fired power plants that would substantially increase emissions of greenhouse gases responsible for global warming. The parties filed, and the court granted, a joint stipulation for dismissal on March 20, 2008.

In *Center for Biological Diversity v. Kempthorne*, an environmental group sued six federal agencies for failing to protect endangered species from the purported effects of global warming.[26] The lawsuit sought a reply from the agencies within 60 days. Specifically, the petition asked the agencies for a review of all threatened, endangered, and candidate species to determine which are imperiled by global warming; a revision of all federal recovery plans to ensure endangered species are able to adapt to a warming environment; a requirement for all federal agencies to implement endangered species recovery plans; a review of the global warming contribution of all federal projects and mitigation of impacts on imperiled species; and technical and financial support to states, local governments, and Native American tribes that voluntarily agree to implement recovery plans.

In addition to environmental group plaintiffs seeking to take litigable action against federal government agencies, there has also been litigation brought by the emitters of GHGs against state government agencies. In *Sunflower Electric Power Corp. v. Kansas Department of Health and Environment*, Sunflower Electric Power Corporation petitioned the court to review a final agency action of the Kansas Department of Health and Environment that prohibited its power plant construction plans.[27] That final agency action was the denial of Sunflower's application for the issuance of a permit authorizing the construction of two new electricity generating units.

There is some communality among environmental plaintiffs in the carbon litigation: the Center for Biological Diversity and the Sierra Club have participated in many of these suits, as well as related litigation challenges not discussed herein, on behalf of plaintiffs. The pace of litigation has picked up after the Supreme Court carbon decision in 2007. No stakeholder—government agencies at the federal, state, and local levels; private manufacturers; electric utilities whether private or public; builders of power plants; or advisory groups to government agencies—is immune from the widening net of litigation involving climate change and global warming. In some instances, feeling aggrieved, the GHGs emitters themselves have initiated litigation as the plaintiff to try to compel actions of government permitting or financing entities. For public companies, this raises issues of disclosure of their carbon-related exposure. Ceres, a group based in Boston, has led in much of this disclosure debate.

With the Obama administration now taking a more proactive policy and legal position on climate change, this diversification of the litigation role of stakeholders in the global warming debate is expected to continue. As regulation of climate change becomes more complex, the focus of litigation will shift from allegations that climate change has not been considered sufficiently, to battles over the specific legislation's constitutionality and compliance with law, and battles over the scope and application of regulations to implement these laws. Key elements of those legal and policy issues are already apparent, and are addressed in the following chapters in this section of the book, as well as in the final section on renewable energy and global warming and adaptation policy.

First, before turning to issues of carbon *leakage* around the contours of existing regulation, the next chapter will examine the new smart grid to limit carbon emissions.

Notes

1. *Massachusetts v. EPA*, 127 1438 (S. Ct. 2007). On April 2, 2007, the United States Supreme Court reversed the United States Court of Appeals for the District of Columbia Circuit. The Court held: (1) that Massachusetts had standing to petition for review, (2) that the EPA has statutory authority under the Clean Air Act to regulate greenhouse gas emissions, and (3) that the EPA acted arbitrarily and capriciously in denying the petition on grounds outside those delineated in the Clean Air Act.

2. See, Ferrey, Steven. 2010. *Environmental Law: Examples and Explanations*. 5th ed. New York: Aspen Publishers. (See chap. 2 for detailed treatment of historic Supreme Court jurisprudence on environmental standing.)

3. *Center for Biological Diversity v. U.S. Dept. of Interior*, F.3d. (D.C. Cir. 2009).

4. *Korsinsky v. EPA*, (U.S. Dist. 2005), LEXIS 21778 (S.D.N.Y. 2005).

5. On September 29, 2005 the judge granted defendants' motion and dismissed the case for lack of jurisdiction.

6. See, *Korsinsky v. EPA*, 192 Fed. Appx. 71–72 (2d Cir. 2006) (discussing standing). Explaining that to establish standing a plaintiff must show that the injury is "'actual' or imminent' rather than 'conjectural' or 'hypothetical.'" Ibid., 71 (quoting *Lujan*, 504 U.S. at 561).

7. *Coke Oven Environmental Task Force v. EPA*, Docket No. 06-1131 (D.C. Cir. 2009). Three environmental groups, 10 states, and two cities originally petitioned the EPA to enact standards regulating greenhouse gas emissions for new stationary sources.

8. *New York v. U.S. EPA*, Docket No. 06-1148 (D.C. Cir. 2009). New York and nine other states petitioned the United States Court of Appeals for the District of Columbia Circuit for review of the decision of the EPA not to adopt stringent emission standards to reduce air pollution from power plants in its Clean Air Act rule released in February 2006.

9. *Center for Biological Diversity, the Sierra Club and San Bernardino Valley Audubon Society v. San Bernardino County*, Docket No. 07-00293 (S. B. Sup. Ct.).

10. *California v. General Motors Corporation*, Docket No. 06-05755 (N.D. Calif. 2007). The State of California filed suit against six automobile manufacturers (General Motors Corporation; Toyota Motor North America, Inc.; Ford Motor Company; Honda North America, Inc.; Chrysler Motors Corporation; and Nissan North America, Inc.). California asserted that the vehicles the defendants manufacture account for 30% of California emissions, and that such emissions, a public nuisance, harm the coastline, water supply, and treasury of California. The automobile manufacturers responded with three major arguments: (1) that the case raised non-judiciable political

questions (i.e., that this is the type of issue for the (political) legislative and executive branches, not the judiciary, to decide) (2) that federal legislation has displaced federal common law on this topic, and (3) that the manufacturers did not cause the injury complained of.

11 *State of Connecticut v. American Electric Power*, 406 F. Supp. 2d 268 (S.D.N.Y. 2005). Plaintiffs were the States of Connecticut, New York, California, Iowa, New Jersey, Rhode Island, Vermont, Wisconsin, and New York City; (NGO) Open Space Institute, and two private land trusts. Defendants were six major power companies: American Electric Power Corp., American Electric Power Service, The Southern Company, Tennessee Valley Authority, Xcel Energy Inc., and Cinergy Corp. The suit claimed that these companies are responsible for 10% of all man-made greenhouse gas emissions in the United States, that these emissions are causing climate change, and that this climate change is harming their sovereign interests as well as those of their citizens. For example, they asserted potential property loss through rising sea levels and public health injuries based on stronger summer heat waves.

12 *California v. GMC*, 47–48 (explaining logistical problem).

13 *Northwest Environmental Defense Center v. Owens Corning Corporation*, Docket No. 04-01727 (D.Ct. Ore. 2006).

14 *Comer v. Murphy Oil USA (Nationwide Mutual Insurance Company)* (a.k.a. the Hurricane Katrina Litigation), Docket No. 05-436 (S.D. Miss. 2009). The plaintiffs filed a class-action complaint for damages and declaratory relief against several oil and coal companies that they claim contributed to climate change through greenhouse gas emissions. They claim that this climate change in turn contributed to the presence and magnitude of Hurricane Katrina and as a result damaged the property of the class.

15 See *Friends of the Earth v. Mosbacher*, 488 F. Supp. 2d at 892 (N.D. Cal. 2007).

16 Ibid., 889. OPIC A.R. Tab 2 at 000014. The handbook provides in relevant part, to provide some degree of [environmental assessment] to every project considered for insurance or finance in determining whether to provide support for the project. OPIC cannot provide a final commitment to a project . . . until its environmental assessment is complete and a determination is made by OPIC that the environmental health and safety impacts of the project are applicable.

17 See, Ferrey, Steven and Anil Cabraal. 2006. *Renewable Power in Developing Countries: Winning the War on Global Warming*. Tulsa: PennWell. Chap. 13.

18 *Inuit Petition to the Inter-American Commission on Human Rights*. 2005. Petition filed before the Inter-American Commission on Human Rights, Washington, D.C. A petition was submitted on behalf of Sheila Watt-Cloutier and 62 other named individuals, all Inuit of the arctic regions of the United States and Canada, claiming human rights violations resulting from the impacts of climate change. They claim that the failure of the United States to reduce greenhouse gas emissions and its refusal to adopt the Kyoto Protocol has resulted in climate change, causing human rights violations toward the Inuit people, including violations of their right to culture and property.

19 *Kivalina v. ExxonMobil Corp.*, Docket No. 08-cv-1138 (N.D. Calif. 2008).

20 *Steadfast Insurance Company v. The AES Corporation*, Docket No. 2008-858 (Va. 2008).

21 *California v. EPA*, Docket No. 07-02024 (D.D.C. 2007).

22 *East Texas Electric Cooperative v. Rural Utilities Service*, Docket No. 2:08-cv-00364 (E.D. Tex. 2008).

23 *Center for Biological Diversity v. Rural Utilities Service*, Docket No. 3:08-cv-01240 (Calif. 2007); 5:2008-cv-00292 (Kent. 2007).

24 *Center for Biological Diversity v. Brennan*, Docket No. 06-07062 (N.D. Calif. 2007); Docket No. 07-16931 (9th Cir. 2007).

25 *Montana Environmental Information Center v. Johanns*, Docket No. 07-01311 (D.D.C. 2007).

26 *Center for Biological Diversity v. Kempthorne*, 466 F.3d 1098 (9th Cir. 2006).

27 *Sunflower Electric Power Corp. v. Kansas Department of Health and Environment*, Docket No. 07-99567-A (Ks. 2007).

12: THE NEW CARBON-ATTUNED SMART GRID: BEYOND SIMPLE POLES AND WIRES

Building a Smarter Grid

New time; new grid. In the United States in 2009, the Obama administration stimulus package included a significant incentive package for the electric sector and the grid, pouring $80 billion in spending and $20 billion in tax incentives into renewable energy and efficiency, as part of the $787 billion stimulus plan.[1] More would be forthcoming in additional transmission system sections of the Waxman-Markey legislation. In the 2009 stimulus package, $4.5 billion is allocated for a better and more reliable delivery system, with most of the money expected to be spent within 18 months principally in the West and Great Plains where there are more renewable power resource developments ongoing. A 30% advanced energy facilities tax credit applies to transmission and grid-related new equipment. Certain transmissions upgrades and extensions qualify for loan guarantees. There is significant focus on the new *smart* grid.

As part of the stimulus package, there is a National Transmission Study to assist constrained renewable resources to reach the market through better transmission and to analyze legal challenges to overcome for a better grid. Certain transmission upgrades and extensions qualify for loan guarantees. It includes $3.25 billion of new borrowing authority each for the Western Area Power Administration and the Bonneville Power Administration to invest in electric transmission grids, entities that already operate 15,000 miles of transmission in the Pacific Northwest. There is also $11 billion for smart grid grants and programs.

President Barack Obama stated that he hoped to see smart meters in 40 million homes, doubling U.S. capacity for renewable energy, and "building a new electricity grid that lays down more than 3,000 miles of transmission lines to convey this new energy from coast to coast."[2] So exactly what is a smart grid? That answer is still evolving. The Energy Independence and Security Act of 2007 required advancement of a smart grid and defines this as follows:[3]

- Use of information and control technology to manage and optimize dynamically the transmission and distribution infrastructure
- Integration of distribution
- Demand response, efficiency, and demand-side resources
- Smart metering technologies to monitor energy use or deploy smart appliances
- Advanced electricity storage and peak-shaving technologies
- Better grid communication

A smart grid, according to the U.S. Department of Energy, provides a digital quality of power and more efficient use of supply resources.[4] The smart grid involves many pieces, but particularly an information and control loop at the delivery point of the grid dividing power delivery into millions of consumer nodes. There are several possibilities of how control will be exercised; however, issues remain as to whether the smart grid is centrally controlled or responds only to end-use consumer action to shape power demand. Power demand peak shaving, electricity storage, and other similar controls are the objective.

In mid-2009, some of these stimulus funds began to flow to grid-related projects. These funds can cover new equipment, software, and communications and control systems, as well as consumer products and distributed generation and energy storage devices. The funds also cover smart meters that measure and store information and can be used for dynamic pricing of power, demand response activities, and remote billing. These funds can finance development of an advanced digital system for enhanced interoperability and cyber security. With more integration of the grid, there is greater potential risk of cyber security breeches that could imperil the large sections of the grid simultaneously.

Also included is a smarter functioning grid to control and optimize the transmission and distribution system, to accommodate distributed generation, new energy storage capacity, or reduction of peak demand.

With more intermittent renewable power expected to be deployed, energy storage (including advanced battery systems, ultra-capacitors, flywheels, compression air energy systems, or other technologies) becomes a critical element of the modern grid. Prevention of overloaded transmission lines, maintenance of transformers, and control of voltage are all targeted goals of this program.

The grid is not just the transmission component of copper wire that carries power at near the speed of light from point A to point B. The grid is the mechanism that conducts power through the interconnected U.S. power network, is dispatched and managed, and thereafter is available to meet electric power requirements in North America. The *grid* is composed not only of approximately 4,800 interconnected power generation resources in the United States, but also of planned more dispersed power generation resources, efficiency capabilities, and self-generation resources. As well, it includes the cable to connect them with consumers, and the human intervention and smart hardware to manage them in an energized instantaneous network. One does not function without the other in a centralized, regional grid, which characterizes a modern power network.

There is much more to the grid than simply wire and poles. It is a constantly replenished energized network capable of doing an infinite variety of work. A constant simultaneous balancing of supply and demand on that system is required. Power moves according to Kirchoff's Law[6] almost at the speed of light on this energized grid.[5] This grid must rebalance supply and demand approximately every four seconds and adjust the amount of supply produced, or risk functional collapse. The importance of the electric sector in global warming abatement is reflected in its changing role. In 1949, only 11% of global warming gases in the United States came from the electric sector; today this share is more than one-third.[7] The Energy Information Administration in 2008 concluded that the electric power sector offered the most cost-effective opportunities to reduce CO_2 emissions, compared to the transportation sector.

Both ends of the grid are in play as it becomes smarter. Future renewable supply sources are less centralized and more diverse than traditional supply sources, whether these new entrants be dispersed renewable generation or onsite cogeneration. The new sources of renewable power are not going to be located where the traditional sources of centralized power are located. Cogeneration also is expected to increase in the new smarter grid.

Cogeneration can use any means of production and is a prime mover for the production of electricity for self generation. Self generation avoids the use of transmission and distribution networks, thus avoiding the more than 50% typical retail charge and various taxes and fees of conventional power supply for the energy user. Cogeneration of electric power and usable heat by facilities on the customer sides of the meters in the new grid is a form of distributed generation and can be more efficient than conventional power generation if they capture more of the energy *in* as usable energy *out* of the process. The total energy produced by a cogeneration system can have much higher efficiency under the first and second laws of thermodynamics. There also can be environmental advantages to use of more of the energy produced.

New grid connections will be required, and a smarter grid will be necessary to balance even more sources in a more complex and intelligent manner. In addition, renewable power introduces an unprecedented degree of intermittency of base power supply to the modern grid. To keep the grid in balance and operational with this new supply reality, there must be the proper mix of new resources not only for primary production of power, but of additional new resources to respond to the constant intermittency of a system more dependent on renewable resources. The importance of this match is underscored by the fact that a slight mismatch in the supply and demand of electric power in California caused brownouts, billions of dollars of extra expense to consumers, collapse of electric sector deregulation, and the recall of the governor.[8]

Managing Greater Grid Reliance on Renewable Power

The challenge with intermittent grid-connected renewable energy

With respect to the grid itself, and how it adapts to an increased renewable power supply share, there are significant issues looming for policymakers and regulators. Primary new supplies of renewable power in the near and intermediate term are wind power and solar power, which are both intermittent in nature. In other words, wind and solar power at a given hour are available in a somewhat unknown duration and strength, intermittently due to natural cycles and forces. Many wind and solar projects expect their resources to be available less than 40% of the hours in an average day.

There are two basic roles power resources can serve from any given source in a centralized grid: baseload power and backup/peaking power. Where in this model's dichotomy will renewable power supply be utilized? Intermittent renewable resources cannot supply reliable baseload power, as they demonstrate a relatively low availability factor often in the 20-40% range of hours during a day or month. Correspondingly, intermittent renewable resources are not of value as reliable backup/peaking power resources, as they are often not available for being called on to fill a need or supplement peak power demand at a particular instant. However, solar power availability typically corresponds to higher demand periods of the day and summer season in the northern latitude.

Given this dichotomy, intermittent renewable resources, such as solar or wind resources, will be operated as part of baseload supply, but their intermittent operation inherently decreases baseload system availability and reliability. Renewable power resources have relatively high capital costs and near zero operating costs because they do not need to purchase any fuel source. What this means is that intermittent renewable resources will be run as much as possible, when available, because their marginal cost of operation, with no fuel costs, is near zero. Most independent system operators, which dispatch and control the regional generation resources, make resource dispatch in the ascending order of lowest cost of operation bid per unit.[9] Therefore, renewable resource units, with zero marginal operating cost, will always be the low-cost first-to-run in place of other more fuel-dependent baseload units.

The North American Electric Reliability Council (NERC) is responsible for maintaining U.S. and Canadian grid reliability. The organization of the North American grid is shown in figure 12-1. According to NERC, carbon regulation and renewable power substitution will compromise grid reliability.[10] There is concern in NERC that the renewable portfolio standards (RPS) in 27 states and 4 Canadian provinces could accelerate early substitution of renewable generators for traditional coal-fired power, and simultaneously decrease grid reliability (see chapter 18 for a discussion of RPS programs). The Waxman-Markey legislation would impose national renewable energy portfolio requirements in addition to existing state legislation. The federal program would not usurp continuation of the 27 state programs, which could continue operating.

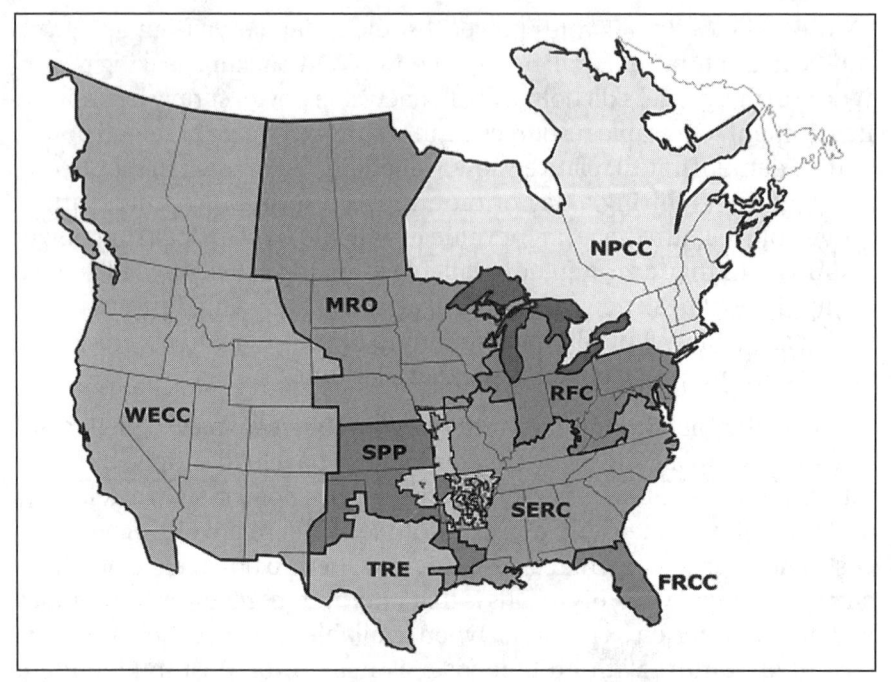

Fig. 12–1. North American Electric Reliability Council regions (Source: NERC)

Traditionally, power planners are concerned about predictable changes in consumer demand creating the need for supply resources to follow changing load (demand). Grid variability can be created from uncontrollable oscillation of electric supply, as well as changing consumer demand. Conventional power generation resources typically fail only because of occasional mechanical difficulty. While this also characterizes renewable resources, renewable resources are much more profoundly affected by regular daily generation unavailability due to natural cycles of wind and sunlight.

An increase in grid-connected renewable power baseload resources, because much of it will be intermittent supply, creates a new daily variable to the grid: renewable intermittent power supply sources can become quickly unavailable. The modern grid will have to constantly, second-by-second, balance supply with demand to keep the grid operational. If power supply does not respond and is deficient vis-à-vis instantaneous demand, the grid can shut down and black out large areas, as happened in the Northeast United States on August 14, 2003.[11]

The reality is that coal-fired or nuclear generation plants, which together constitute the backbone of U.S. and many other countries' baseload electricity generation supply, once turned off even temporarily to make room for renewable power resources, cannot be quickly brought back online. They require very long multi-hour or day-long warm-up periods to restart, are not capable of quick-start response or capability when needed, and cannot follow or respond to changes in electricity load or power supply.

This increased reliance on intermittent resources can reduce the reliability of the power grid unless there are advancements in power storage technology compared to what is now available. Energy storage in the United States is equivalent to about 2.7% of generating capacity. With decreased system resource availability and reliability, there will be more demand for quick-start backup power generation resources to fill the greater volatility of baseload resources.

The critical role of new quick-start backup power resources in the new grid

What does baseload intermittency do to the U.S. power system? First, it changes how units are dispatched (operated), and the need for new backup power generation technology to fill the voids created by supply intermittency. Economist Brent Bartlett modeled all 4,800 existing power projects in the United States and how they will respond to U.S. carbon allowance auction. His modeling indicates that what happens with auction of carbon allowances is that certain high-carbon baseload resources, especially those powered by coal, are forced out of the dispatch queue by increasing cost factors. In their places, certain existing gas-fired backup generation units are pressed to operate more, due to their lower requirement for auctioned carbon allowances and thus lower total operating cost, and are used more for baseload and backup power. Second, because of this phenomenon, a carbon regulating system requires more, rather than less, backup power resources in the grid, at the same time that it actually converts and diminishes the number of economic existing backup power generation resources available to the grid in this mode. This represents a bit of a disconnect between existing technologies and policies.

Moreover, the current grid configuration in the United States and in other industrialized countries already features a significant shortfall of existing backup power resources, and particularly backup resources that are either capable of operating on dual-fuel inputs and/or have quick-start capability to be available on short (10-minute) notice. Each of these factors will prove critical in a period where both availability of sufficient fossil fuel resources and the pricing of fossil fuels have been volatile and unreliable over recent periods.

The New England grid control area provides an interesting example of these phenomena. With about 31,052 MW of rated generating capacity to serve a peak demand of 28,127 MW, this doesn't afford the recommended 15–20% surplus for equipment repairs and unit unavailability at peak times.[12] However, the peak power demand has been increasing over time as a percentage of average demand. In 1980, New England peak capacity was 154% of average load, and increased in 1990 to 159% and in 2000 to 175%.[13] This is a function of increasing air-conditioning usage during the summer peak days. The peak is forecast to continue to increase over time.[14] New York City, for example, has a peak demand almost twice its average load.[15]

The need for peaking power resources in New England is established as 7,000 MW.[16] However, only 1,510 MW of non-pumped storage-peaking resources are available.[17] With pumped storage counted, there are about 3,000 MW of peak power resources. This is only 5–10% of total electric power supply, now even before the more assertive rollout of renewable resources. This is more than a 50% deficiency between peak power need and supply.[18] Moreover, these limited available peaking power resources are largely oil-fueled when there is a need in today's volatile fuel market for dual fuel capability: only 20% of this already inadequate peak power resource in New England has dual fuel oil/gas capability.[19] Two-thirds of the remaining 80% of the peaking power is generated by oil fuel only.[20] Oil is more polluting and thus more responsible for CO_2 global warming emissions per unit of power generated than is natural gas.

Therefore, the existing backup/peaking capacity is dramatically short of where it needs to be even though the system has enough power generation resources in gross. This is compounded by lack of either dual-fuel or less polluting gas-fuel alternatives. The grid operator for New England, ISO-NE, analyzing this situation concluded, "A lack of fast-start resources in transmission-constrained subareas could require the ISO to use more costly resources to provide these necessary services. In the worst case, reliability could be degraded."[21]

What is important in an age of renewable power and carbon control is quick-start capability of sufficient backup/peaking electric supply resources. Most of the existing backup/peaking capacity now installed in the grid is not the newer aero-derivative quick-start technology. Quick-start allows the shutdown generator to go from a cold start to full power production in less than 10 minutes. Therefore, it is almost instantaneously available without having to be *spinning* and operating prior to need, just to be ready in case, as many current backup generators do at considerable financial and environmental expense.

Conventional non-aero-derivative generators take an extended period of hours to bring their temperatures up gradually from a cold start, and similarly must ramp down their temperatures slowly.[22] This means that conventional backup/peaking fossil fuel-fired units continue to burn fossil fuel, and thus to pollute, just to get ready to provide peaking power when/if later needed. These spinning reserve units also expel a much less contained profile of environmental emissions when operating in a spinning mode at partial operation levels trying to ramp up to be available. Moreover, while spinning to increase their temperatures to their design values so as to be ready, the power that these units could or do produce may or may not be used by the grid and that power can incur power *uplift* costs to transmit on the grid. This multiple loss is incurred by the grid and its ultimate power consumers whether or not these units are ever required to supply backup power when the peak needs actually arrive. This is already the state of affairs in many regions before the expected rollout of renewable resources.

Maintaining required grid reliability

The emergence of wind generation is a reality: wind power dominates renewable power choices. There were record wind installations in 2008, in excess of 8,000 MW, or 42% of all new generation additions. With more reliance on solar and wind power—when unexpectedly the wind does not blow or the sun is occluded by clouds, and thus renewable power generation units are not available—there is no ability to quickly start most conventional backup power units.

The impact of this on existing systems is already manifest. On February 26, 2008, the Electric Reliability Council of Texas (ERCOT) grid operator, which is a leader in wind power deployment, was unable to compensate with sufficient backup power resources when there was an unexpected drop in forecast wind power production by more than 80%.[23] This was in part due to inadequate planning, misinterpretation of system data, and failure of certain backup units to respond as expected. However, regardless of cause, the grid has had instances already when it was unable to respond as needed without requiring consumers to shed load and compromise reliability.

So, as one attempts to transform the grid to accommodate an unprecedented increase in renewable power, the power generation grid in many regions is short of needed backup/peaking power resources. What backup resources they do have are not quick-start or suited to serve a grid utilizing a fast-climbing fraction of intermittent renewable resources. With a rollout of more intermittent renewable resources, conventional backup/peaking generation resources with long multi-hour warm-up periods are not available to quick-start on unpredictable short notice. A new type and deployment of backup/peaking power generation resources will be needed to allow the grid to efficiently accommodate greater renewable resources.

The very good news is that the quick-start technology is available, demonstrated, and cost-effective. There are no technological barriers to overcome. Therefore, the grid will need to accompany more renewable resources with a whole new battalion of quick-start peaking power resources to fill in their not entirely predictable intermittent daily operation. How the costs of this requirement will be assigned have not yet been resolved. Are these costs to be borne by the renewable power generators whose lack of reliability will increase backup requirements for the grid? Transmission imbalance penalties can be imposed by transmission owners on certain intermittent power sources, such as intermittent renewable projects. Or are these costs to be borne by all consumers and ratepayers of power? In either case, a major change in the nature of the grid will be required as the country moves to more use of wind and solar power, with their inherent intermittency due to natural forces, and the current lack of cost-effective power storage.

There are suggestions that demand response resources might be able to fill this gap when solar resources are not available. In the 2008 ISO-NE forward capacity auction, the forward capacity market new demand response resources totaled 1,188 MW and existing demand response resources totaled 1,366 MW. These demand response bids total almost 10% of total current peak load.

It is generally concluded that energy efficiency is available at a cost of about $0.03/kWh saved by the efficiency investment.[24] The demand-side management (DSM) possibilities between 2007 and 2010 are estimated to be over 230 TWh, or equivalent to about 5.5% of the forecast electricity power requirements in 2010.[25] This total DSM potential could trim 7.5% of peak period electric consumption. The North American Electric Reliability Council estimates that interruptible load and direct control load management reduces national summer peak by about 2.5%. A 2009 report to the Federal Energy Regulatory Commission (FERC) concluded that with smart meters communicating the real-time pricing of power, 37–188 GW of peak power demand could be cut. This represents a cut of up to 20% of the 782 GW peak 2007 U.S. electric demand.

FERC's report on demand-side response and metering states that through its regulation of regional transmission organizations and independent system operators, FERC can ensure comparable treatment of demand response resources in ancillary service markets, and allow these resources to bid into the organized energy market and reflect these contributions of lost load during an operating reserve shortage.[26] The Energy Policy Act of 2005 required that within 18 months, electric utilities shall offer certain customer classes a time-based rate schedule, which options included time-of-use pricing, critical peak pricing, and real-time pricing.

What is equally important is that the American grid capture wasted dispersed energy sources occurring on the customer side of the meters. For example, industry expels as waste heat a significant fraction of its energy use. By capturing that waste heat before it exits the stack into the ambient environment, and converting it to electric power, there can be a substantial dispersed creation of power for onsite use or into the grid.[27] This changes the basic flow of power on this grid, but it is an essential part of an efficient and smart grid, to capture as usable energy what is now exhausted as waste. New load control software allows the capability to control building management systems remotely, capture real-time energy data, and accurately compute customer baselines of energy use.

Transmission Infrastructure Extension to Support a More Sustainable Power System

Connecting new resources

Many new renewable power resources are located a long transmission distance from the load that uses that power. The transmission grid in the United States is relatively old. As an example, the New England grid has been criticized for now engaging in $11 billion in annual trades of electricity over wires built approximately 40 years before to serve a much more limited number (about one-third) of players in a tightly regulated utility environment.[28] Without new stimulus, transmission only represents about 10% of the rate base of expenditures by U.S. utilities. With the aging of the transmission system, efficiency of that system decreases. In 1970, average line loss was 5%; in 2001, average line loss increased to 9.5% as lines, transformers, and circuit breakers of the transmission system aged. Aging of the grid is not an asset.

However, again here, the good news is that these challenges are technologically resolvable. While there can be extended controversy among interest groups in siting transmission infrastructure, these are political and legal disputes, and not technical. And these issues of transmission infrastructure have been present before when U.S. utilities after World War II chose to construct large baseload facilities often located a distance from load centers. Large transmission infrastructure had to be created to move this power between source and consumer. However, from technical and legal perspectives, this poses challenges for the existing power grid.

Issues also concern the cost of new transmission infrastructure to reach the location of some wind and solar installations in more remote locations. For wind, facilities must be sited where there is a good wind regime. This often is not in densely populated load centers.

While renewable resources are distributed across the United States and the world, they are not distributed evenly.[29] Nine states east of the Mississippi River do not have any subregions with very high wind resources.Six states from Virginia to Massachusetts do not have any subregions with at least one-quarter billion metric tons of currently available biomass annually. These northeastern regions of the United States have relatively dense populations and significant electricity demand. While they have access to renewable resources, those renewable resources are not as concentrated as in other areas of the United States. However, with many buildings in densely populated areas, there is the potential of tapping energy efficiency as a substitute for additional generation capacity.

Transmission infrastructure must be constructed to bring renewable power from the generation source to the load center. Who pays for this transmission infrastructure is an issue. There is debate in various regions of the country about whether the generating resource should pay for the cost of its power interconnect, whether those who choose to buy that renewable or other particular power should pay for getting that power to them, or whether the costs should be socialized by spreading the transmission cost among all ratepayers in a state, a region, or nationally. By socializing the transmission costs, the real costs of building certain high-cost power generation resources, especially at hard-to-reach remote locations, are masked in providing this power to load centers. Texas allows cost recovery through rate base for transmission connections within Competitive Renewable Energy Zones.[30] California offers special cost sharing for transmission in "locationally constrained" areas.[31] This includes rate base recovery.

Establishing specific grid corridors

An increase in use of renewable energy will require new transmission corridors and capacities to transport that power from the generation site to the load centers. There are several things that can be done to improve delivery operating capabilities of existing and new grid infrastructure. First, dealing with the operating efficiency of the transport function of the grid itself, monitoring frequency, voltage, and control areas can be switched to monitoring phase angles of output of the electric current wave.[32] Synchrophasor technology lets the grid continuously monitor several times each second what is happening on each phase of the system and correct irregularities. Second, grid operators can control an increasing percentage of load remotely. Third, more distributed generation can supplement new capacity that is controlled by the grid.

Multistate regional planning for transmission facilities will be necessary. At issue is whether renewable resources should have their own renewable-only transmission corridors, or whether they should share general transmission corridors with all supply sources. These decisions have financial implications, which FERC may not be able to wholly shape. The U.S. Court of Appeals for the Fourth Circuit in 2009 denied FERC's claim that it had authority to override a state denial of transmission infrastructure construction permission pursuant to the Energy Policy Act of 2005.[33] That FERC authority could be restrengthened under the Waxman-Markey legislation.

Massive new transmission infrastructure is planned to bring externally generated power into the carbon-regulated states. The largest of these projects is the American Electric Power (AEP) Interstate Project, which would put in place a 765 kV transmission line stretching from West Virginia to New Jersey. Other examples include the Trans-Allegheny Interstate Line (TrAIL) Project, being undertaken by Allegheny Power to enhance transmission capability from western Pennsylvania to Maryland and Virginia, and the Meadow Brook to Loudon 500 kV line proposed by Dominion Resources to carry power into the Washington D.C. metropolitan area. Recently, Oncor and others were selected for a $5 billion transmission project to connect future wind farms in West Texas to the load centers in metropolitan areas.[34] Western governors have asked the federal government to pay for transmission extensions to reach areas where renewable energy projects might be built, as part of economic stimulus efforts.

Southern California Edison Company, alone, is looking to spend more than $5 billion on transmission projects between 2008 and 2013 in order to add about 7,000 MW of grid-connected renewable generation to its system. The California Public Utilities Commission allowed Southern California Edison to spend $4.5 million of ratepayer money to participate in identifying renewable resource zones and developing transmission plans to access resources placed in those zones to deliver power to load centers. These zones would tend to be in Nevada, Arizona, and Southern California.

Some conflicts are emerging between states. The Arizona Corporation Commission rejected Southern California Edison's proposal to build a 230-mile line to provide Southern California with access to cheaper Arizona power, fearing that the exported power would increase costs to Arizona consumers who enjoyed the benefits of cheap existing plant output. The U.S. Supreme Court in 1982 held that harboring inexpensive wholesale power in-state was a violation of the dormant Commerce Clause of the U.S. Constitution. Refusing to allow the transmission of power out of state, by not approving the construction of necessary transmission capacity within the state, accomplishes indirectly what the Supreme Court forbade states to do by regulation. Texas utilities are also spending a similar sum as Southern California Edison to bring Texas competitive renewable energy resources to market.[35]

Massachusetts regulators have shown skepticism about paying for new interconnections and a power line to Maine, which would allow transport of wind power south to load centers. The new transmission line from Maine to load centers in southern New England states is opposed by the Maine Public Advocate Office.[36] Maine utilities also have requested adders to their base return on equity for transmission facilities to move new renewable power from northern Maine.[37] There are ongoing disputes about whether new capacity on transmission lines must be made available on a competitive open access basis. Traditionally, the interconnection from the independent power producer (IPP), whether renewable or not, to the existing transmission lines, has been the responsibility of the IPP to construct and finance.

Wind and solar grid-connected power projects do not use transmission capacity of the grid efficiently, given their intermittency. A study released in 2008 by Cambridge Energy Research Associates found that the production patterns of wind farms "do not correlate well with peak summer demand," and "capacity provided by wind projects is typically valued at 10–20% of their maximum rated capacity."[38] In other words, for satisfying demand requirements of the grid, wind and solar resource intermittency debase their grid capacity value.

The very positive news is that there is still time to deal with all of this. Even with a stimulus package, renewable portfolio standards, and climate legislation, the move to a substantial quantity of renewable power will take time, which will allow for new smart grid construction and extension. Although the push into renewable and efficiency may be relatively vigorous, it will come in smaller increments than more concentrated conventional fossil fuel-fired or nuclear power projects. Each energy supply project, whether conventional or renewable, requires siting and permitting, which involves process and time. By definition, the rollout of renewable and energy efficiency projects will not be immediate in impact. Therefore, there is time to adapt the utility grid and supply to these changes as they are deployed.

The U.S. Department of Energy forecast in 2008 that the United States could achieve 20% of its electricity from wind power by 2030.[39] This possibility is actually much more aggressive than is likely under current scenarios, but is technologically feasible. The Green Communities Act, enacted in Massachusetts in 2008, sets goals to achieve 15% of energy through renewable resources, an additional 5% of energy supply through alternative energy (defined as not traditionally renewable), and 25% of electric load through demand-side management, by 2020.[40] It also sought to have certain renewable resources be onsite generation resources, instead of remote grid-connected resources. This puts different demands on an aging grid that now connects large centralized plants in a few locations to consumers.

Decoupling and Grid Operator Incentives

From a regulatory perspective, new ways of regulating transmission providers to decouple their rates and earnings exclusively from the total volume of power distributed (to reflect various rate recovery mechanisms tied to explicit policy incentives) is gaining support. Decoupling the revenue stream determination of regulated distribution utilities from the volume of power they sell is a critical reform, with several states trying to provide incentives for great efficiency in energy supply.

FERC reported at the end of 2008, that 10 states had adopted polices to decouple changes in utility revenue from changes in utility sales volume.[41] California, New York, New Jersey, Maryland, and Massachusetts are the five states leading decoupling. These are the states that led electric utility restructuring and retail deregulation a decade ago, the development of renewable portfolio standards and renewable system benefit charges over the past decade (see chapter 18), and have led in state carbon regulation over the past year (see chapters 7–9).[42] These states have some of the highest consumer retail electric prices in the United States.

Massachusetts, in the middle of 2008, passed new legislation that will dramatically compel the rollout of renewable energy technologies.[43] Certain renewable energy owners are urging regulatory commissions to allow utilities to sign long-term contracts with power suppliers. Massachusetts also is supporting such long-term contracts with renewable energy suppliers to be entered by utilities. A bill in Rhode Island would allow such a utility to earn an incentive profit of 3% of annual contract payments under such a long-term contract, in addition to its normal rate of return.[44] In addition, the Rhode Island legislation would eliminate stranded costs for the purchasing utility by allowing it to immediately resell such long-term renewable power in the wholesale spot market.

There also has been decoupling pressure at the federal level that has cleaved some nongovernmental organizations (NGOs). Various environmental groups in 2009 urged Congress, as part of economic stimulus legislation, to provide incentives or enticements for states to decouple electric utility revenues from utility sales volume. Originally, there was a revenue decoupling requirement in the proposed 2009 stimulus proposal for states to delink utility rate of return determinations for the volume of power sales, to garner competitively awarded funds. It was dropped in the version enacted so that it only required an indication that the state is moving in that direction.

This move of environmental groups split (1) consumer groups, state regulators, and some industrial groups, which are concerned about increasing costs through various incentives, from (2) environmental groups that want to provide more conservation incentives to utilities. This split between environmental and consumer groups, over similar issues, also occurred in the last great flurry of new federal energy legislation during the energy crises in the late 1970s. "It's consumers versus utilities and environmentalists," according to one observer. States are just starting to feel their ways along this new path. Yet, it is a critical component of changing regulatory incentives for power system operation.

So adapting the new grid not only requires adapting the architecture of copper wire to connect more dispersed renewable and other generating sources to load centers, but also to develop an alternative suite of backup and peaking generating sources to fill the more intermittent profile of the new grid when wind and solar resources provide a larger share of the power. While at one level this is an issue of new hardware, it is accompanied by regulatory issues of who pays for the significant cost of this new architecture. It is a challenge where the regulatory and legal challenges are at least as vexing as the engineering rollout.

The next chapter turns to issues of great recent attention:

- Leakage of carbon emissions across regulated borders
- The ability of states to regulate outside of their states
- The state efforts to stop leakage pursuant to the Commerce Clause of the U.S. Constitution

Notes

1. U.S. Congress. 2009. American Recovery and Reinvestment Act of 2009. Public Law 111–5, 111th Cong., 123 Stat. 115.

2. Tiernan, Tom and Jeff Ryser. 2009. Revised Language in House Bill Eases Fears on Smart Grid Provisions, but Concerns Linger. *Electric Utility Week*. February 2: 1, 34; Cash, Cathy et al. 2009. Senate Pushing for Big-Picture Grid Plans; 'Shovel-Ready' Projects a Question. 2009. *Electric Utility Week*. February 2: 1, 34.

3. See, Energy Independence and Security Act of 2007. Public Law 110–140, 110th Cong., 42 U.S.C. sec. 17381-86.

4. See, United States Department of Energy. 2008. *The Smart Grid: An Introduction*. 10–13. http://www.oe.energy.gov/DocumentsandMedia/DOE_SG_Book_Single_Pages(1).pdf. This document describes the smart grid.

5. This law is also called Kirchhoff's first law, Kirchhoff's point rule, Kirchhoff's junction rule, and Kirchhoff's first rule. The principle of conservation of electric charge is that at any point in an electrical circuit where charge density is not changing in time, the sum of currents flowing toward that point is equal to the sum of currents flowing away from that point.

6. See, Ferrey, Steven. 2004. Inverting Choice of Law in the Wired Universe: Thermodynamics, Mass and Energy. *William and Mary Law Review*. (April): 1,839.

7. See, Energy Information Administration, http://www.eia.doe.gov/oiaf/1605/ggrpt/excel/historical_co2.xls.

8. Ferrey, Steven. 2004. Soft Paths, Hard Choices: Environmental Lessons in the Aftermath of California's Electric Deregulation Debacle. *Virginia Environmental Law Journal*. 23(2): 251.

9. For a discussion of independent system operators, see Ferrey, Steven. 2010. *The Law of Independent Power*. Eagan, Minnesota: Thomson/West Publishing. Sec. 10:87.

10 Carbon Control News. 2008. Public Utilities Fear that GHG Cuts Might Threaten Electricity Supply, Reliability. *www.carboncontrolnews.com*. July 28.

11 Wald, Matthew, Richard Perez-Pena, and Neela Banerjee. 2003. The Blackout: What Went Wrong; Experts Asking Why Problems Spread So Far. *New York Times*. August 16: A1. This article examines the cause of the 2003 blackout across the northeastern United States.

12 See, www.iso-ne.com; *Application of Montgomery Billerica Energy Partners, L.P.*, Massachusetts EFSB 07-02, at p. 3-12.

13 *Application of Montgomery Billerica Energy Partners, L.P.*, Massachusetts EFSB 07-02, at Figure 3.3-2.

14 Ibid. See figure 3.3-3; ISO New England Inc. 2006. *Capacity Energy Loads and Transmission Forecast Report 2006–2015*. April.

15 Wood, Lisa. 2009. New York Readies for Stimulus Funds with Order to Utilities on Metering Pilots. *Electric Utility Week*. February 16, 33.

16 See, Massachusetts Energy Facilities Siting Board. 2008. *In the Matter of Braintree Electric Light Department Petition*. Case No. EFSB 07-1/D.T.E/D.P.U. 07-5, Final Order. February 29: 78.

17 Massachusetts Energy Facilities Siting Board. 2009. *Order, In the Matter of Montgomery Energy Billerica Power Partners*. Docket 07-02. March.

18 *Application of Montgomery Billerica Energy Partners, L.P.*, 3-8 and 3-16.

19 Massachusetts Energy Facilities Siting Board, *Braintree Order*, 78; a quadrupling of dual-fuel-fired backup/peaking capability is immediately needed. See, ISO New England Inc. 2005. *Regional System Plan 2005*. ES-2.

20 Ibid; only 260 MW of peaking capacity in New England has dual-fuel capability. See, ISO New England Inc., *CELT Report 2006–2015*.

21 ISO New England, 2006. *Regional System Plan*, 5.

22 U.S. Department of Energy. 2006. *Supplement to the Draft Environmental Impact Statement for the Gilberton Coal-To-Clean-Fuels And Power Project*. December: 3–4; citing CO_2 Capture and Storage Working Group 2002, CO_2 Capture and Storage in Geologic Formations, NCCTI Energy Technologies Group, Office of Fossil Energy, U.S. Department of Energy, January 8, 2002. http://www.netl.doe.gov/publications/carbon_seq/CS-NCCTIwhitepaper.pdf.

23 Electricity Journal. 2008. How Renewables Can Be Undermined by Intermittency. *Electricity Journal*, June: 5.

24 Environmental Protection Agency. 2005. *Energy Efficiency Action Plan – Context and Framework*. 12. www.epa.gov/solar/documents/leadership-meeting-05/eeap_pre.pdf. (presuming $0.03/kWh).

25 Gellings, C. et al. 2006. Assessment of U.S. Electric End-Use Energy Efficiency Potential. *Electricity Journal*. November. This citation quotes the Keystone Institute 2003 report and 2006 Annual Energy Outlook of U.S. Department of Energy, Energy Information Administration.

26 U.S. Federal Energy Regulatory Commission. 2008. *2008 Assessment of Demand Response and Advanced Metering*. Staff Report, i-ii.

27 Casten, Thomas R. and Phillip F. Schewe. 2008. Getting the Most from Energy: Recycling Waste Heat Can Keep Carbon from Going Sky High. *American Scientist*. http://www.americanscientist.org/search/home_result.aspx?q=Casten%20and%20Schewe.

28 Platts. 2008. New England Grid Is On Borrowed Time; Groups Warn It Will Soon Exceed Limits. *Electric Utility Week*. January 14: 1, 23. The report charges that transmission inadequacy has already resulted in approximately $2 million in extra charges to consumers since 2003. Approximately 70% of U.S. transmission lines and transformers are at least 25 years old, and 60% of circuit breakers are more than 30 years old.

29 Kutscher, Charles, ed. 2007. *Tackling Climate Change in the U.S.: Overview and Summary of the Studies*. American Solar Energy Society.

30 See, Nowamooz, Alborz. 2008. Inadequacy of Transmission Lines: A Major Barrier to the Development of Renewable Energy. *Environmental and Energy Law and Policy Journal*. 3: 176. This article discusses the Texas plan.

31 See, California Independent System Operator. 2008. *2008 Summer Loads and Resources Operations Preparedness Assessment*. 9. This publication notes the connection of remote resources.

32 See, Ferrey, Steven. 2000. *The New Rules: A Guide to Electric Market Regulation*. Tulsa: PennWell. 12–13.

33 *Piedmont Environmental Council v. FERC* (4th Cir. 2009) WL 388237, No. 07-1651.

34 Carr, Houseley. 2009. Oncor, AEP/MidAm, LCRA, Others Are Selected for $4.92 Buildout of Texas Transmission. *Electric Utility Week*. February 2: 3.

35 Power Markets Week. 2008. Transmission Boom Calls for Reconsidering Cost Allocation Methods, Some Officials Say. *Power Markets Week*. July 28: 11. This article reports the Texas Public Utility Commission spending $5 billion on transmission costs.

36 Wood, Lisa. 2009. Solar Company Proposes Unique Project to Displace Need for 350 MW Maine Line. *Electric Utility Week*. February 2: 10–11.

37 See, FERC Docket No. EL08-77. http://www.ferc.gov/docs-filing/dec-not.asp.

38 Ryser, Jeffrey. 2008. With Wind Power at Their Back, 13,000 at Conference Weigh Pros, Cons. *Electric Utility Week*. June 9: 1, 32.

39 United States Department of Energy. *20% Wind Energy by 2030: Increasing Wind Energy's Contribution to U.S. Electricity Supply*. http://www1.eere.energy.gov/windandhydro/pdfs/41869.pdf.

40 See, Green Communities Act. 2008. Massachusetts Acts 50.

41 FERC. *2008 Assessment of Demand Response and Advanced Metering*, ii.

42 See, Ferrey, *The New Rules*, chap. 8.

43 See, Green Communities Act.

44 See, Rhode Island General Laws. 2008. Sec. 39-26.

13 CARBON LEAKAGE AND THE COMMERCE CLAUSE

Leakage and Policy Choices

The *what* and *where* of leakage

A major practical and policy problem identified by some European countries (see chapter 10), the RGGI states,[1] as well as California,[2] is so-called carbon leakage. Leakage occurs when "generators outside of the capped region export power to load-serving entities within the region without being covered by the regional carbon cap,"[3] and is defined within the RGGI regulatory scheme "as the increase in CO_2 emissions outside the RGGI region that may 'net out' (or partially eliminate) a portion of the emissions reductions made within the RGGI region under the Program."[4]

The reality is that leakage will occur as CO_2-producing activities that are regulated and limited under a particular region's carbon control program move outside that region so as not to be regulated, thereby eliminating net reductions in emissions with the shift of economic activities' locations. Developers in non-carbon-regulated states will have economic incentives to build and/or operate CO_2-emitting facilities where they do not have to incur the cost of acquiring carbon allowances and complying with regulations. In such an event, carbon costs will be borne, to their competitive disadvantage through higher settled power prices, by competing facilities in the carbon-regulated region.

This results in negating the environmental improvements that would otherwise result from the carbon-reduction program.[5] For example, since

RGGI includes 10 discreet East Coast states, the immediately proximate power plants outside the RGGI region are to the west and south—or upwind in terms of migration of power plant emissions generally from west to east. Figure 13-1 illustrates those states regulating carbon and those proximate states from which leakage could occur into the regulated areas over the carbon transmission lines that connect the country. A shift to out-of-region generation is a shift to additional upwind pollutants from more operation of external polluting power plants, producing electricity that otherwise would have been generated by cleaner RGGI-region compliant projects.

WHO ARE THE CLIMATE LEADERS?
Building a clean energy economy with cap and trade

MGGRA: Midwest Greenhouse Gas Reduction Accord
RGGI: Regional Greenhouse Gas Initiative
WCI: Western Climate Initiative

The states and provinces in green are leading the fight against global warming by shifting away from fossil fuels. They are putting a legal limit on carbon pollution (the cap) and designing market-based programs to reduce pollution economically (the trade). The climate leaders shown here represent more than half of North America's population and economic activity.

Sightline
INSTITUTE
www.sightline.org

Fig. 13-1. States that regulate carbon still face leakage issues from neighboring states that do currently impose carbon restrictions. *(Source: Sightline Institute, 2008)*

The results of modeling commissioned by the RGGI Staff Working Group found that a substantial proportion of CO_2 emissions avoided by RGGI will be offset by corresponding increases in non-RGGI states.

Leakage from neighboring states like Pennsylvania is a significant concern under RGGI. RGGI is projected to have a significant leakage problem even if CO_2 allowances sell for only the relatively modest long-term price of $7 per ton of CO_2. To date, allowance costs have remained below that threshold.

The early modeling showed leakage as high as 90% depending on the programmatic assumptions. The final models predict annual leakage into the RGGI states of external CO_2 between 40% and 57% over the life of the RGGI program. Moreover, this 2006 modeling was conducted without assuming Maryland's participation in RGGI, and Maryland's current participation is predicted to provide an additional conduit for leakage. Provisions of RGGI state regulations that would permit allowances to be purchased and retired by non-affected facilities will also increase leakage. Thus, leakage is a significant concern for the RGGI program.

An additional issue affecting leakage involves regional power generation reliability. In its 2008 Reliability Needs Assessment, the New York State Independent System Operator (NYISO) analyzed the effects of RGGI on energy supply reliability in the control region. It concluded that because there will be a finite number of carbon allowances, RGGI creates the risk that generators could not obtain sufficient allowances "to meet bulk power system electricity needs and also comply with the RGGI program."[6] NYISO concluded that RGGI could further threaten reliability through allowance hoarding, allowance market manipulation, or the removal of allowances from the market (e.g., by purchasers that are not using allowances for RGGI compliance). Because New York, like other regional areas, must have certain specific plants operating to satisfy electric demand at certain times, if allowances were not available, plants could not operate and electric reliability would be compromised. Under RGGI, there is no ability to pay a fine for a shortfall in possessed allowances; civil and criminal penalties under the Clean Air Act Title V operating permit are instead imposed.

This leakage threat is very real to the goals of regional carbon control initiatives. In trying to decrease the amount of CO_2 emissions from RGGI states by 55 million tons over the period of 2009–2019, an increase of unregulated power imports from carbon-uncapped coal-fired plants in states such as Ohio and Pennsylvania of even 1.5% to 2.5% could wipe out all scheduled emissions reductions from regulated generators within the regulated carbon RGGI region. These minor increases in leaked power would negate program benefits.

Transmission additions now proposed to the grid (see chapter 12) increase the risk of leakage. There are multibillion-dollar projects to build electric transmission infrastructure that would allow electricity generated by high-carbon emission coal-fired power plants to travel east into the RGGI region. RGGI states such as New Jersey, New York, Maryland, and Delaware are bordered by states that are not signatories to RGGI and historically produce a large of volume of electricity from coal-fired power plants.[7] All are interconnected by high-voltage transmission lines. Similarly, California imports power from 11 states served through the Western Area Power Administration, including a large amount of coal-fired power.

Concern about leakage is growing. Massive new transmission infrastructure is planned to bring externally generated power into the carbon-regulated states. The largest and most significant of these projects is the American Electric Power (AEP) Interstate Project, which would put in place a 765 kV transmission line stretching from West Virginia to New Jersey. There are other similar projects on the drawing boards. The Trans-Allegheny Interstate Line (TrAIL) Project, being undertaken by Allegheny Power to enhance transmission capability from western Pennsylvania to Maryland and Virginia, and the Meadow Brook to Loudon 500 kV line proposed by Dominion Resources to carry that power into the Washington, D.C. metropolitan area, are other examples.

The policy response

In the initial phase of RGGI program design and at the time the RGGI guiding principles agreement was drafted, leakage as a separate economic phenomenon was not a consideration. The extent of leakage predicted by subsequent modeling, however, was projected to dramatically diminish the benefits and cost-effectiveness of the program. Also of note, subsequent state RGGI proposals abandoned the traditional method of direct allocation of allowances to affected facilities without cost in favor of the currently proposed 100% allowance auction design, which could increase leakage (see chapter 14 for more on auction).

RGGI state efforts to bar the RGGI border to leakage are already under way. The governors in affected RGGI states agreed to "pursue technically sound measures to prevent leakage from undermining the integrity of the [p]rogram."[8] Note that the concern is something that works technically, although legal considerations are equally important. The group's memorandum of understanding provides for a multi-state RGGI Emissions Leakage Staff Working Group to "consider potential options for addressing leakage" and to issue a report assessing those options.

To stem this inflow of power from outside the RGGI control region, the RGGI states have discussed implementing some type of control, regulation, or tax to discourage cheaper power imports to load-serving entities (LSEs) from unregulated states external to the RGGI regions. Such regulation by the RGGI states will have to target power flows based on their state of power generation origin, distinguishing between those from RGGI states and non-RGGI states. Such controls on the free flow of electricity from other states, where electricity is a commodity or service that is a quintessential article in interstate commerce, run up against the so-called dormant Commerce Clause of the U.S. Constitution.

The New Jersey energy regulator was required to develop a plan to reduce leakage of power into that state. Public Service Electric & Gas Company (PSEG) proposed a plan to have New Jersey curtail imports of high-carbon power from out-of-state cheaper power supplies by requiring regulated retailers to purchase a certain amount of power from RGGI-covered suppliers.[9] Even at a modest auction price of $6.35 per ton for RGGI allowances, the rationale is that leakage into the state would stop RGGI's goal of fostering low-carbon power by increasing the import of less expensive high-carbon power into the state by 26% to 35%. The New Jersey Department of the Public Advocate responded that it would require a subsidy of about $50 per MWh to gas plants to make state gas-fired plants as cost-efficient as out-of-state coal-fired plants.

Such responses would enact a form of regulation that discriminates relative to power sales based on geographic point of origin of the power. New Jersey state legislation prohibits energy efficiency measures from being deployed to mitigate potential leakage, unless other methods are found to violate the Constitution. It thus favors regulation of conduct rather than incentives for demand-side management (DSM) alternatives or conservation. The Public Advocate criticized another proposal for New Jersey to extend the state's RGGI cap to cover imported generation, as creating Commerce Clause violations.

The assault against leakage by the early states is ultimately a fight of *us* (a state regulating carbon from its power generators) versus *them* (neighboring states or foreign countries that do not similarly regulate carbon emissions from their power sectors). In Europe, these leakage concerns can pit one country's concerns against bordering countries (see chapter 6). This is a very real policy concern because the import of higher-carbon power was already occurring prior to carbon regulation; it may now accelerate. In the end, leakage could increase total regional carbon emissions and emissions by out-of-state producers—counter to the intended policy goal to limit carbon, which has a global impact.

Slowly, the legal issue emerged: A March 2008 RGGI working group report urged states to be cautious in trying to tax or adopt measures to frustrate leakage from outside the RGGI region. Such legal mechanisms used to control such leakage are geographically based discriminatory regulation. In the U.S. legal system, this immediately raises dormant Commerce Clause concerns, and invokes the most exacting "strict scrutiny" legal standard of court review, under which few similar state regulations have survived.

U.S. Commerce Clause Constitutional Requirements

The dormant Commerce Clause

The specific mechanism for structuring and protecting state or RGGI carbon regulations must not run afoul of constitutional Commerce Clause requirements. Traditionally, a state's ability to protect its citizens' "health, life, and safety" is a valid exercise of its power.[10] The retail regulation of utilities is also a traditional function of local police power of the states.[11] The generation and transmission of electric energy, however, are activities particularly likely to affect more than one state.

The Commerce Clause provides that "[t]he Congress shall have Power [t]o regulate Commerce...among the several States."[12] The Supreme Court has recognized that "it is difficult to conceive of a more basic element of interstate commerce than electric energy, a product used in virtually every home and every commercial or manufacturing facility."[13] Although the Commerce Clause is an affirmative grant of power, the Supreme Court has also interpreted it as limiting the states' ability to "unjustifiably...discriminate against or burden the interstate flow of articles of commerce."[14] Under the Federal Power Act of 1935, the federal government exercises regulatory power over the wholesale power market, while the states are left alone to regulate most retail transactions.[15]

Although the Commerce Clause affirmatively grants Congress the ability to regulate interstate commerce, there is no clear directive limiting states' abilities to regulate where Congress has remained silent. Nevertheless, the U.S. Supreme Court has consistently held that the Commerce Clause exerts a prohibitive force limiting states' powers to regulate interstate commerce in certain situations, even where Congress has not regulated.[16] Therefore, although states are permitted to promote in-state businesses, they are not permitted to protect those businesses from out-of-state competition by enacting laws that "benefit in-state economic interests by burdening out-of-state competitors."[17]

The power of the Commerce Clause "has long been understood to have a 'negative' aspect that denies the States the power unjustifiably to discriminate against or burden the interstate flow of articles of commerce."[18] This specific aspect of the "dormant" Commerce Clause has been interpreted by courts as a tool used to prevent the states from splitting into separate economic entities.[19] The issue of the application of the dormant Commerce Clause has been characterized as the oldest question in constitutional law.

Standards and different outcomes

There are two standards of judicial review under a dormant Commerce Clause challenge to state regulation. First, geographically based discrimination is evaluated under a strict scrutiny test applied by the federal courts, and such a geographic statute, with rare exceptions, is found to be *per se* a violation of the dormant Commerce Clause. Second, state regulation that is not geographically based, but nonetheless has incidental impacts on interstate commerce, is evaluated under a balancing test where the state interest is weighed against the degree of impairment of interstate commerce. Moderately discriminatory state actions can in some circumstances survive this challenge.

The construction of the dormant Commerce Clause is one of the most litigated environmental and energy issues before the Supreme Court in the last quarter century. Where a regulation *clearly* on its face discriminates against interstate commerce or has that practical effect, that regulation violates the Constitution unless some justification for the discrimination unrelated to protectionism is demonstrated.[20] Point of origin discrimination regarding articles in commerce to protect in-state interests to the detriment of interstate commerce "is *per se* invalid," unless the state can identify a legitimate and compelling local interest that can be served by no other means.[21] If local regulations discriminate facially or by intent against interstate commerce based on geographic location, whether by regulation or taxation, courts apply a "strict scrutiny" standard and there is a high probability that the regulation will be invalidated.[22] Any statute or regulation that facially discriminates against interstate commerce by giving "differential treatment [to] in-state and out-of-state economic interests that benefits the former and burdens the latter" will be "virtually *per se* invalid."[23]

On the other hand, there are cases in which a state may be exercising traditionally recognized authority (including protection of health, environment, natural resources, and safety), and not discriminating based on geographic locus, but the effect nonetheless is to discriminate against the free flow of interstate commerce. In such cases, the court will balance the interest of the state against the burden on commerce, and will evaluate less offensive means of effectuating the purpose of local regulation.[24] A nondiscriminatory regulation supported by a legitimate state interest, incidentally burdening interstate commerce, can be constitutional unless the burden on interstate commerce is clearly excessive in relation to the local benefits.

With such a balancing it is not necessary to demonstrate that the state statute is necessarily the least restrictive means to accomplish the stated purpose:[25] "[N]ondiscriminatory regulations that have only incidental effects on interstate commerce are valid 'unless the burden imposed on such commerce is clearly excessive in relation to the putative local benefits.'"[26] These are two distinct standards for judicial review. Challenged environmental regulations that have been deemed valid under the less stringent "incidental burden" standard have balanced the "substantial state interest in promoting conservation of energy and other natural resources" against any burdens on interstate commerce.[27]

Because no bright line separates regulation that does and does not discriminate, and the judicial standard applied by the court is so distinct between the two, the critical determinants in analyzing a challenged regulation are the court's initial conclusion as to whether or not a regulation is discriminatory,[28] and if so, whether such discrimination is based on geographic point-of-origin regulation.[29] If a regulation is discriminatory and the discrimination is based on point-of-origin regulation, judicial strict scrutiny will apply. Even in the absence of a discriminatory intent, courts are able to find Commerce Clause violations and strike state regulations to prevent "balkanization" that would result in an inconsistent patchwork of regulations.[30]

The Key Legal Comparison on Carbon Controls

The combination of (1) a tax or charge on exterior supplies, with (2) a subsidy to certain in-state activities, as is now in place for some of the RGGI states, factually and legally parallels the U.S. Supreme Court decision in *West Lynn Creamery, Inc. v. Healy*.[31] The Court there found a violation of the dormant Commerce Clause in the state regulatory scheme. It conceded that either part of the program considered alone—the tax or the payments—would probably be constitutional. However, the Court assessed the "entire program," declining to "divorce the premium payments from the use to which the payments [were] put." The scheme imposed the net burden of the tax on out-of-state producers.

What is important to note is that many of the arguments the State of Massachusetts advanced to the Supreme Court to defend that regulation also apply to carbon regulation, all of which the Supreme Court entirely rejected as mitigating elements. First, the State in *Healy* argued that because its pricing regulation or tax was applied only to in-state transactions, it was "nondiscriminatory." In other words, a state should be allowed to tax or penalize its own in-state transactions. "Even gross receipts derived from sales of services to be performed wholly in one State are taxable by that State." Nonetheless, the Court found that the combination of tax on interstate articles and subsidized projects together violated the Commerce Clause.

Second, the State argued that since the direct subsidization of domestic industry is *per se* constitutional, the combination of tax and subsidy—each allowed in its own right—would not violate the dormant Commerce Clause. The Court disagreed:

> A pure subsidy funded out of general revenue ordinarily imposes no burden on interstate commerce, but merely assists local business. The pricing order in this case, however, is funded principally from taxes on the sale of milk produced in other States.... [W]hen a nondiscriminatory tax is coupled with a subsidy to one of the groups hurt by the tax, a State's political processes can no longer be relied upon to prevent legislative abuse, because one of the in-state interests which would otherwise lobby against the tax has been mollified by the subsidy.[32]

The Supreme Court focused on how combined tax and subsidy schemes undercut normal political checks and balances to find that they violated constitutional principles.

Third, the State argued that because the milk dealers who incurred the charges in *Healy* were wholesalers, and thus not direct competitors of the Massachusetts dairy farmers who received the awarded subsidies and who were producers, the scheme imposed no discriminatory burden on commerce. The Supreme Court rejected this argument by holding that "the imposition of a differential burden on any part of the stream of commerce—from wholesaler to retailer to consumer—is invalid, because a burden placed at any point will result in a disadvantage to the out-of-state producer."[33]

With the RGGI carbon scheme, even this previously unsuccessful argument could not be advanced, as the restricted or taxed out-of-region electricity wholesalers are direct competitors of the in-region electricity power wholesalers. In fact, most of the RGGI states have restructured their regulatory systems and partially deregulated their electric sectors to promote exactly such wholesale competition in power supply in a constant daily competitive market. Of the 10 participating RGGI carbon-regulating states, all but Vermont have restructured their electricity sectors in the past decade. So the RGGI carbon scheme lacks even some of the factual insulation that was still unsuccessful in *Healy*.

Fourth, the State argued that any incidental burden on interstate commerce resulting from the pricing order in *Healy* is outweighed by local benefits, including "protecting unique open space and related benefits."[34] The environmental argument is one that has been raised in most of the commerce clause legal adjudications of environmental regulation that have occupied the courts. The Court states that "even if environmental preservation were the central purpose of the pricing order, that would not be sufficient to uphold a discriminatory regulation."[35] The use of facially discriminatory economic means taints an otherwise laudable end and violates the dormant Commerce Clause.

On balance, the Court found Massachusetts' pricing charge and subsidy regulatory scheme to be "clearly unconstitutional" because "[i]ts avowed purpose and its undisputed effect are to enable higher cost Massachusetts dairy farmers to compete with lower cost dairy farmers in other States." Any state surcharge or tax on out-of-region carbon-emitting wholesale power imports bears many similarities to this scheme. The arguments that the states or groups of states regulating carbon will advance to defend regulation will parallel those advanced by Massachusetts to defend its regulation of milk in the *Healy* matter.

Are There Any Legally Key Distinctions with Carbon?

With climate control, the carbon regulation at issue is of a different article in commerce than the milk in *Healy*. Carbon is expelled generally into the environment, rather than sold as a packaged commodity, and is directly related to the production, flow, and sale of power, an article in interstate commerce. With carbon, the proposed state schemes would create an emission right, rather than a market in a commodity *per se*. Yet that emission right has an intimate connection to a quintessential article of interstate commerce: electric power.

Can the regulation of greenhouse gases be legally or factually distinguished from the regulation of milk with respect to the Commerce Clause so as to allow it to survive? Although there are significant factual distinctions between milk and power distribution, the constitutional distinction between them is scant in terms of carving out a legal exception to distinguish carbon regulation under the dormant Commerce Clause. The following sections examine briefly a few such possible distinctions, and whether they rise to the level of legal significance regarding climate control.

Selective nature of subsidies between carbon and other products

In *Healy*, approximately one-third of the milk tax was imposed on milk in-state, and two-thirds on milk in interstate commerce from out-of-state producers. So, the two-thirds out-of-state wholesalers subsidized the one-third of the producers located in-state. In some of the electricity deregulated states, less than two-thirds of the taxed power may come from out of state. But whether the balance of power taxed from out of state is one-third or two-thirds does not alter the basic legal distinction based on the point of origin of the power. The Court in *Healy* focused on the ultimate economic impact, noting that "the [assessment is] effectively a tax which makes milk produced out of State more expensive. Although the tax also applies to milk produced in Massachusetts, its effect on Massachusetts producers is entirely . . . offset by the subsidy provided exclusively to Massachusetts dairy farmers."[36] A surcharge and subsidy system of any size was held by the Supreme Court as still discriminatory based on point of origin.

In versus out: market segmentation and energy services

Is there a meaningful regulatory distinction between the subject matters of carbon and milk, or between dairy goods and electric services? While milk is a classic good, states vary as to whether electricity is considered a good or a service, which could influence whether it is deemed an article in interstate commerce.[37] However, emissions from the production of power certainly cause air emissions that are capable of interstate movement, and will have a value in commerce as carbon emissions are monetized and traded.

While the trend at FERC appears to be to consider electricity a good,[38] and state law is mixed,[39] it should not fundamentally alter the Commerce

Clause analysis, although no court has examined whether power plant emissions are articles in commerce. As a matter of basic physics, a moving electron is an electron, no matter where it originates. Because CO_2 emissions are directly linked to interstate power flows, the key question here is whether the carbon regulatory scheme is based on point of origin of the power and carbon; if so, it is evaluated under a strict scrutiny test which presumes *per se* invalidity, subject to very few exceptions.

A surcharge on, or barrier to, higher-carbon, out-of-state wholesale power will also not pass muster. While the state can segment the market among resources, a state cannot discriminate in price or require an arbitrary price be paid for wholesale power. In *New England Power Co. v. New Hampshire*, the Supreme Court overturned a New Hampshire Public Utilities Commission regulation that restricted the export of privately owned hydroelectric energy produced within the state by a multi-state wholesale company.[40] The New Hampshire regulation of in-state hydroelectric power, based exclusively on its point of origin, attempted to reserve cheaper hydroelectric power for consumers within the state. The Supreme Court held such discriminatory pricing of interstate power through state regulation to be facially discriminatory and a violation of the dormant Commerce Clause, in spite of the states' traditional power to regulate the retail electric market. It is also clear that regulation in one state cannot affect the actual physical flow of power from or in another state.

It is not relevant in Commerce Clause jurisprudence whether the particular state regulation attempts to keep commerce in or out of the region.[41] Power is a quintessential article in interstate commerce, moving almost at the speed of light. Therefore, the distinction of trying to keep certain preferential power in the state (such as in the *New Hampshire* case) or keep certain non-preferred power out of the region (in carbon regulating states, for example) does not create a direction of regulation that is legally significant. With carbon regulation, if a state or region attempts to segregate certain power within or without the region based on its point of origin, it creates similar constitutional concerns.

Environmental protection rationales

With milk, an environmental rationale also was raised to defend the milk regulation, including "protecting unique open space and related benefits." The *Healy* Court did not accept the rationale. In *City of Philadelphia v. New Jersey*, the Court held that however legitimate a state's ultimate environmental protection purpose, such may not be accomplished by discriminating against out-of-state articles of commerce, unless justified by some rationale apart from place of origin.[42] The Court in *Healy* found that even if environmental preservation were the central purpose of the milk pricing order, that would not be sufficient justification to uphold a discriminatory regulation. The Court has consistently maintained that a Commerce Clause violation occurs from either discriminatory purpose or discriminatory effect—either by the design or application of regulation.[43]

Role of the wholesale ISO

Does the supposed deregulation of power markets in many states, and the creation of regional independent system operators (ISOs) to control the transmission system, change the analysis? FERC has encouraged the formation of ISOs and of regional transmission organizations (RTOs). However, the price for the sale of wholesale power is still regulated by FERC.

The RGGI states and California operate their wholesale regional power markets in similar ways. Non-utility generators may sell power to a customer or a retailer through the regional ISO power market. Each participant is required to submit hourly balanced generation and demand bids to the ISO. Taking these balanced bids, the ISO controls the dispatch of generation, manages the reliability of the transmission grid, provides open access to transmission facilities, and provides ancillary services (see chapter 12).

Most of the power purchased by providers in a carbon-regulating state will have previously passed through a wholesale transaction to arrive eventually at the retail user. The entirety of this wholesale power flow is in interstate commerce and is subject to terms and conditions set by FERC, rather than state jurisdiction. There is no residual state jurisdiction of any kind over wholesale power transactions, let alone their restriction (see chapter 14).

Possible Saving Doctrines under the Commerce Clause

Compensatory tax doctrine

The Court has addressed and overturned a number of interstate taxation schemes, including transfer taxes[44] and a first-use mineral/energy tax.[45] Note that some states are imposing the obligation to purchase auctioned carbon emission allowances on the first sellers of power (see chapter 14).

Although the Supreme Court has deemed that a state statute being discriminatory "may be a fatal defect," it has also outlined an exception known as the compensatory tax doctrine, which "justif[ies] a facially discriminatory tax as achieving a legitimate local purpose that cannot be achieved through nondiscriminatory means."[46] To qualify, such a tax must "impose on interstate commerce the rough equivalent of an identifiable and 'substantially similar' tax on intrastate commerce." As a threshold matter, the state must identify the intrastate burden for which it is attempting to compensate. Once identified, this tax must be shown to "approximate—but not exceed—the amount of the tax on intrastate commerce." Finally, "events on which the interstate and intrastate taxes are imposed must be 'substantially equivalent' . . . serv[ing] as mutually exclusive prox[ies] for each other." The challenge in assigning any type of surcharge or emissions cap-and-trade program on imported electricity into those states that regulate carbon would be establishing a uniform way to measure both the emissions from the in-state regulated generators and the out-of-state non-regulated generators to satisfy the "equivalent burdens" aspect of the exception.

Quarantine exception

Another recognized exception to invalidity under strict scrutiny of geographically based regulatory discrimination is the necessary quarantine of commodities or things. The narrow quarantine exception was recognized in *Maine v. Taylor*.[47] The state demonstrated in that case that there was a distinct danger to the ecosystem and no less discriminatory way to realize the state interest than quarantining the commodity out-of-state. Even in applying this exception, the Supreme Court stated that the "Commerce Clause significantly limits the ability of States and localities to regulate or otherwise burden the flow of interstate commerce, but it does not elevate free trade above all other values."

Problems exist, however, in extending the holding of *Maine v. Taylor* to a carbon-regulating scheme for out-of-state electricity. First, in contrast to a key fact in *Taylor*, in-state CO_2 is identical to out-of-region CO_2. Second, CO_2 impact is not local and isolated as the environmental toxin in *Taylor*; it is cumulatively an international problem. Third, there are less discriminatory alternatives to achieve CO_2 reductions available, some of which have even been proposed by committees associated with RGGI. Therefore, it is hard to legally transfigure carbon emission regulation into the judicially limited last-resort quarantine exception recognized in *Taylor*.[48]

Proprietary actions

There is one additional exception where, in a proprietary mode, a state may marshal and control its *own* energy resources, even if that discriminates in favor of in-state interests, and against out-of-state interests or interstate commerce. In a proprietary mode, a state can burden "commerce which would not exist if [the state] had not decided to subsidize a portion of the . . . business."[49] Recently, in *United Haulers Association, Inc. v. Oneida-Herkimer Solid Waste Management Authority*, the Supreme Court determined that a local "flow control" ordinance that required that waste from a particular area go solely to one waste-handling facility was valid.[50] The key distinction was that the regulated entity was publicly owned. That is not the case with carbon regulation under the RGGI or the California schemes, however.

Moreover, there is precedent, originating in the RGGI region, regarding regulation of privately owned power differentially based on state of origin. In New Hampshire, an effort to allocate less expensive in-state hydroelectric power for local consumers, as opposed to more expensive out-of-state power produced from fossil fuels by the same multi-state utility, was stricken by the Supreme Court.[51] State carbon regulation does not entail quarantining an unavoidable toxin; instead, such regulation allows copious quantities to be emitted, but charges for that right or implements modest restrictions. Moreover, CO_2 is not a toxin; it is essential to life on Earth. Therefore, the proprietary exception and the quarantine exception would not apply.

Here, none of the three exceptions—compensatory tax, quarantine, or proprietary—should apply to state regulation of CO_2 emissions from their privately owned power sector generators.

States are now considering how to secure their borders to prevent subsidizing leaked high-carbon power across state lines. So as states attempt to stop the very real concern about leakage, they may discriminate in regulation between in-region carbon-regulated power, and the import of out-of-region power not carbon-regulated. In doing so, they can violate provisions of the Commerce Clause, which as part of the U.S. Constitution is designed to prohibit such point of origin regulatory discrimination against interstate commerce.

There is a second constitutional concern with state carbon regulation: it involves the wholly new choice to auction carbon emission allowances, and how that is examined under the constitutional Supremacy Clause and the filed rate doctrine. The next chapter will discuss that constitutional concern.

Notes

1 Regional Greenhouse Gas Initiative. 2005. *Memorandum of Understanding in Brief.* 9–10. http://www.rggi.org/docs/mou_brief_12_20_05.pdf.

2 Cap and Trade Subgroup. 2006. *Cap and Trade Program Design Options.* California Climate Action Team. http://www.climatechange.ca.gov/climate_action_team/reports/2006report/2006-03-27_CAP_AND_TRADE.PDF.

3 Cowart, Richard. 2006. *Regulatory Assistance Project, Addressing Leakage in a Cap-and-Trade System.* 1. http://www.raponline.org/Pubs/RC-Leakage-4-06.pdf.

4 New York State Department of Environmental Conservation. 2008. *Final Generic Environmental Impact Statement.* 107. http://www.dec.ny.gov/docs/air_pdf/rggigeis.pdf.

5 See, Regional Greenhouse Gas Initiative. 2007. *Stakeholder Meeting Stakeholder Comments.* March 12: 17–18. www.mass.gov/dep/air/climate/rggisum2.doc.

6 New York Independent System Operator. 2007. *2008 Reliability Needs Assessment, Supporting Documents, and List of Appendices for the 2008 Comprehensive Reliability Planning Process.* http://www.nyiso.com/public/webdocs/services/planning/reliability_assessments/2008_RNA__Supporting_FINAL_REPORT_12_12.pdf.

7 RGGI Emission Leakage Multistate Working Group. Potential Emissions Leakage and the Regional Greenhouse Gas Initiative. March 2007: ES-1.

8 Regional Greenhouse Gas Initiative, *Memorandum of Understanding*, 10.

9 Powers, Mary. 2008. PSEG 'Leakage' Plan Would Cost New Jersey Ratepayers $50 Million Annually, Group Says. *Electric Utility Week.* July 14: 8.

10 *Huron Portland Cement Co. v. City of Detroit*, 362 U.S. 440, 443 (1960).

11 *Arkansas Electric Cooperative Corp. v. Arkansas Public Service Commission*, 461 U.S. 375, 377 (1983). Case cites *Munn v. Illinois*, 94 U.S. 113 (1977).

12 U.S. Constitution. Art. I, sec. 8, cl. 3.

13 *Federal Energy Regulatory Commission v. Mississippi*, 456 U.S. 742, 757 (1982).

14 *Oregon Waste Systems, Inc. v. Department of Environmental Quality of Oregon (1994)*, 511 U.S. 93, 98; *H.P. Hood & Sons, Inc. v. DuMond*, 336 U.S. 525, 534–38 (1949); *Cooley v. Board of Wardens*, 53 U.S. (12 How.) 299 (1851).

15 See, Title II of the Public Utility Act of 1935. 74 Public Law 333, 49 Stat. 803, 838–863. This amends the Federal Power Act.

16 See, *Reading Railroad Co. v. Pennsylvania*, 82 U.S. (15 Wall.) 232, 279–80 (1872); *Brown v. Maryland*, 25 U.S. (12 Wheat.) 419, 448–49 (1827); *Gibbons v. Ogden*, 22 U.S. (9 Wheat.) 1, 208 (1824).

17 *New Energy Co. of Indiana v. Limbach*, 486 U.S. 269, 273 (1988) [citing *Bacchus Imports, Ltd. v. Dias*, 468 U.S. 263, 270–273 (1984); *H.P. Hood & Sons v. DuMond*, 336 U.S. 525, 532–533 (1949)].

18 *Oregon Waste Systems, Inc. v. Department of Environmental Quality of Oregon*, 511 U.S. 98 (1994). This case cites *Wyoming v. Oklahoma*, 502 U.S. 437 (1992); *Welton v. Missouri*, 91 U.S. 275 (1876).

19 See, *Hughes v. Oklahoma*, 441 U.S. 322, 325-26 (1979). This case notes that the Commerce Clause is intended to prevent "economic balkanization" among the states, and recognizing that the dormant Commerce Clause limits state regulation as much as federal regulation of commerce.

20 *C. & A. Carbone v. Town of Clarkstown*, 511 U.S. 383, 402 (1994) [citing *Wyoming v. Oklahoma*, 502 U.S. 437, 454 (1992); *Maine v. Taylor*, 477 U.S. 131, 138 (1986)].

21 *Carbone v. Clarkstown*, 392 (citing *Maine v. Taylor*, 138).

22 See, *H.P. Hood & Sons, Inc. v. DuMond*; *City of Philadelphia v. New Jersey*, 437 U.S. 617 (1978). This case found a ban on interstate waste disposal in private facility impermissible; *West v. Kansas Natural Gas Co.*, 221 U.S. 229 (1911). This case held that an attempt by a state to prohibit export of natural gas discriminated against interstate commerce on such basis.

23 *Oregon Waste Systems v. DOEQ*, 93, 99.

24 See, *Hughes v. Oklahoma*; *City of Philadelphia*; *Dean Milk Co. v. City of Madison, Wisconsin*, 340 U.S. 349 (1951). This case held that the city must use a less discriminatory means of regulating the quality of milk sold and choose a nondiscriminatory method to effectuate such purpose.

25 See, *Minnesota v. Clover Leaf Creamery Co.*, 449 U.S. 456 (1981). This case held that a Minnesota statute banning plastic milk containers as environmentally unacceptable served a legitimate purpose and was sustained, notwithstanding the fact that it promoted local industry at the expense of out-of-state industry. *Pike v. Bruce Church, Inc.*, 397 U.S. 137 (1970). This case held that the state may not restrict fruit packaging because the restriction was clearly imposed to protect local industry and burdened interstate commerce.

26 *Oregon Waste Systems*, 99 (quoting *Pike v. Bruce Church*, 142).

27 *Minnesota v. Clover Leaf Creamery*, 473.

28 See, *Hunt v. Washington State Apple Advertising Commission*, 432 U.S. 333 (1977).

29 *City of Philadelphia v. New Jersey*, 437 U.S. 617 (1978); *West Lynn Creamery, Inc. v. Healy*, 512 U.S. 186 (1994).

30. See, *Hughes v. Oklahoma* (invalidating a statute that placed no limit on the number of minnows that could be taken by licensed minnow dealers but forbade any person from leaving the state with more than three dozen minnows).

31. *West Lynn Creamery, Inc. v. Healy*.

32. Ibid., 199–200.

33. Ibid., citing *Brown v. Maryland*, 25 U.S. 419, (12 Wheat.) 419, 444, 448 (1827).

34. Ibid., 207 n. 20 (quoting Brief for Respondent at 40).

35. *West Lynn Creamery, Inc. v. Healy*, 512 U.S. 207 n. 20 (quoting *City of Philadelphia v. New Jersey*, 437 U.S. 626–27). "[W]hatever New Jersey's ultimate purpose, it may not be accomplished by discriminating against articles of commerce coming from outside the State unless there is some reason, apart from their origin, to treat them differently."

36. *West Lynn Creamery, Inc. v. Healy*, 195.

37. For a discussion of the goods versus services distinction for electricity, see Ferrey, Steven. 2000. *The New Rules: A Guide to Electric Market Regulation*. Tulsa: PennWell. Chap.12; Ferrey, Steven. 2004. Inverting Choice of Law in the Wired Universe: Thermodynamics, Mass and Energy. *William & Mary Law Review*. 45: 1865–1888.

38. In several cases, FERC applied the U.C.C. rule or an equivalent, thus reinforcing its determination that electricity is a "good" for the purposes of law. See Commonwealth Elec. Co., 46 FERC par. 61,253, par. 61,759 n.15 (1989); Village of Jackson Center, 91 FERC par. 63,013, par. 65,123–124 (2000); Cent. Ill. Pub. Serv. Co., 20 FERC par. 61,043, par. 61,089 (1982); Golden Spread Elec. Coop., 40 FERC par. 61,348, par. 62,047 (1987).

39. See, Ferrey, Steven. 2004. Inverting Choice of Law in the Wired Universe: Thermodynamics, Mass and Energy. *William & Mary Law Review*. 45: 1865–1888.

40. *New England Power Co. v. New Hampshire*, 455 U.S. 331 (1982).

41. See, *City of Philadelphia v. New Jersey, Fort Gratiot Sanitary Landfill, Inc. v. Michigan Department of Natural Resources*, 504 U.S. 353 (1992), *Oregon Waste Systems, Inc. v. Department of Environmental Quality*, and *Chemical Waste Management, Inc. v. Hunt*, 504 U.S. 334 (1992). These cases concern keeping articles of commerce out of the region. Compare previously mentioned cases with *C. & A. Carbone, Inc. v. Town of Clarkstown* and *U. & I. Sanitation v. City of Columbus*, 205 F.3d 1063 (8th Cir. 2000). These cases concern keeping articles of commerce in the region.

42. *City of Philadelphia v. New Jersey*, 617, 626–27. See also, *Carbone, Inc., v. Clarkstown*, 383, 393. This case held that the town cannot "justify the flow-control ordinance as a way to steer solid waste away from out-of-town disposal sites that it might deem harmful to the environment. To do so would extend the town's police power beyond its jurisdictional bounds."

43. *Bacchus Imports, Ltd. v. Dias*, 263, 270; *Minnesota v. Clover Leaf Creamery*, 456, 471 n. 15.

44 *Boston Stock Exchange v. State Tax Commission*, 429 U.S. 318 (1977). This incentive offended the Commerce Clause by "foreclos[ing] tax-neutral decisions." Ibid., 331.

45 *Maryland v. Louisiana*, 451 U.S. 725 (1981). The Louisiana legislature enacted a tax of $0.07 per thousand cubic feet on the "first use" of any natural gas imported into Louisiana that was not previously taxed by another state or the federal government. Ibid., 731–32. The statute defined "first use" as selling, transporting, processing, treating, using in manufacturing, or "other ascertainable action." Ibid.

46 *Oregon Waste Systems, Inc. v. Department of Environmental Quality of Oregon*, 511 U.S. 93 (1994).

47 *Maine v. Taylor*.

48 See, *Dean Milk Co. v. City of Madison* (holding that a state must first pursue reasonable nondiscriminatory regulatory alternatives); *Fort Gratiot Sanitary Landfill, Inc. v. Michigan Department of Natural Resources*.

49 *Hughes v. Alexandria Scrap Corp.*, 426 U.S. 794, 815 (1976); (Stevens, J., concurring).

50 *United Haulers Association, Inc. v. Oneida-Herkimer Solid Waste Management Authority*, 127 S. Ct. at 1790, 1798 (2007).

51 See, *New England Power Co. v. New Hampshire*, 455 U.S. 331 (1982).

14 CARBON ALLOWANCE AUCTION: REGULATORY AND LEGAL ISSUES

Federal Preemption of State Carbon Regulation

Even putting aside regulatory mechanisms to address possible leakage that triggers dormant Commerce Clause problems (see chapter 13), the U.S. state carbon regulatory schemes trigger other constitutional issues that may be examined under the doctrine of federal preemption. This is a function of how state regulators choose to implement state carbon regulation programs by electing to auction allowances necessary to operate power plants, rather than provide them as a free allocation as per all other historical allowance regimes. Terry Tamminen, an energy advisor to California Governor Arnold Schwarzenegger, stated that the "potential legal challenges could pose the biggest stumbling block" to California's climate change initiatives.[1]

Motive and Program Design: the Auction

The mechanics of auction power pricing

As discussed earlier (see chapter 7), the RGGI states have veered off the original program agreement for RGGI whereby allowances to emit carbon would be allocated without charge to existing emission sources. The guiding principles agreement provided that "[t]he initial phase of the cap and trade program will entail the allocation and trading of carbon dioxide allowances to and by sources in the power sector only." However, the 10 states now implementing RGGI utilize an auction of available allowances to the highest bidder, not an allocation to affected facilities requiring allowances as contemplated in the guiding principles agreement.

By auctioning 100% of allowances, the carbon schemes are intentionally designed to impose higher costs (via the requirement to purchase allowances) on certain high-carbon coal- and oil-fired power plants. This will change the trading price of all wholesale power in the region. Three control regions of the RGGI states set power prices: New York Independent System Operator, Independent System Operator New England, and Pennsylvania-Jersey-Maryland Independent System Operator (PJM). In California, the California Independent System Operator (Cal-ISO) sets power prices. Because all of these ISOs work through a second-price auction where the highest accepted bid price for wholesale power supply for a given hour determines the price of all power (including lower bid power),[2] the marginal, highest cost unit sets the price for all power.[3] That highest-cost unit for some hours of the day will be the high-carbon units that have to buy more allowances per kilowatt-hour of power produced than other units. However, this higher price will be earned through the second-price auction by all wholesaling power suppliers under the controlling regulations, and the resulting price shock will reverberate through all power sales.

This reality is not reflected by the regulating states. Taking the RGGI scheme as an example, the auction of 100% of the allowances is defended by New York regulators as not changing the cost of the program, despite the added cost of purchasing needed allowances:

> A plant operator will include the value of the allowances needed to operate in its electricity price because the value represents the opportunity cost of using the allowances to operate rather than selling the allowances on the market. Given this dynamic, allocating allowances to generators at no cost is not cost effective. Furthermore, the cost of the Program does not increase if the generators are required to purchase the allowances, because the generator incorporates the same dollar value of the allowance in its bid to supply electricity whether the allowance was obtained at no cost or through purchase on the open market. Allocation ensures that the value of the allowances is used to promote the emissions reduction goals of the program through cost-effective energy efficiency and clean energy technologies, while simultaneously reducing the cost of the Program to consumers.[4]

However, because the most expensive marginal unit that sets the hourly clearing price for electric energy is almost always an affected regulated facility (i.e., subject to RGGI carbon allowance purchase obligations), the cost of RGGI carbon allowances will be reflected in the clearing price that is the highest offer accepted for service.[5] This clearing price will uniformly be paid ultimately by consumers to every power plant that is dispatched during the hour. The marginal cost of the most expensive power purchased determines the price paid for all power at each hour, with the second price auction system employed in the control regions of each of the RGGI states and in California. Those facilities that do not have to purchase allowances in order to operate and sell wholesale power (e.g., unregulated out-of-state carbon-emitting power plants selling power into an RGGI-regulated state) will thus collect a price for their wholesale power reflecting the cost of RGGI allowances embedded in the highest hourly clearing price paid for power, even though they themselves have no RGGI compliance costs.

Consequently, certain power generating facilities, including higher carbon-emitting, out-of-state facilities, will receive increased revenues due to the implementation of RGGI or similar regulatory schemes.[6] Moreover, it is conceivable that the relative cost advantage provided to out-of-state generating facilities due to the ISO-determined power purchase scheme could result in investments for new electric generating supplies being channeled to cheaper-to-operate non-RGGI states at the expense of the RGGI state economies. This is the problem of leakage (see chapter 13).

The auction battles in the states

Fights over the use of proceeds from the auction of carbon allowances have already emerged in the RGGI scheme. The Connecticut governor sought to funnel some of these proceeds back to state ratepayers to offset higher power prices and was halted by an opinion of the Connecticut attorney general that doing so would violate the state's RGGI legislation that dedicates such funds to energy efficiency and renewable energy.[7] Lurking behind these issues is the constitutional question of whether selling allowances and dedicating and utilizing proceeds, either for in-state energy efficiency investments or in-state consumer rebates, raises any Commerce Clause issues (see chapter 13).

California's joint agency staff paper on allocation of GHG allowances in the electric sector recommended a combination of allocating and auctioning emission rights.[8] The California Air Resources Board (CARB) favored a gradual approach to auctioning allowances, with the auction percentage increasing over time. In January 2009, a CARB Advisory Board recommended that all allowances be auctioned rather than dispersed without charge. Major fights have erupted in California over the allocation and auction of CO_2 emission allowances. One major battle is whether allowances will be dispersed without charge to load-serving entities, and if so, whether the traditional load served or the traditional level of emissions should constitute the basis for distribution. The California investor-owned utilities in May 2008 submitted comments to regulators, urging California to allocate carbon allowances to all emission sources based on historical power output, rather than emissions output, employing a uniform GHG baseline. This would favor the award of allowances to less carbon-intensive sources and utilities.

Surplus allowances could be sold. Dynegy and other independent power providers in California that operate higher-carbon electricity generators believe that allowances should be distributed based on historic emissions levels, rather than power output, to "recognize the reliability benefits conferred by such sources," and the "loss of market value of these resources."[9] However, environmental groups charge that any allocation based on historic emissions results in "grandfathering," which "rewards historical polluters, penalizes early actors, could lead to windfall profits, and asks the biggest polluters to reduce their emissions the least."[10]

The concept of auctioning carbon allowances and capturing substantial payments has been extremely attractive to carbon regulators across the United States. Seven western states participating in the Western Climate Initiative also recommended that 25% to 75% of total emission allowances be auctioned in their own proposed regional market design (see chapter 9). But several state regulators, including those from California and Washington, have also acknowledged that the states could be legally preempted in their efforts to regulate carbon.

The stated motive for auctioning carbon emission allowances

Environmental officials in the various carbon-regulating states have declared that the rationale for the auction of 100% of the carbon allowances is to increase the cost of carbon-emitting power generation and capture of profits as state revenues. The leader of RGGI, the New York Department of Environmental Conservation (NYDEC), issued public statements claiming that the decision and purpose of the auction of 100% of carbon allocations is to prevent affected electric generators from earning "excess" profits resulting from the operation of the wholesale market.[11] Environmental officials went on record as implementing the auction of all allowances to reduce the rate of return that high-carbon power generators receive pursuant to their FERC-approved market rates, which NYDEC considers to include "excess" profits. Thus, the RGGI auction policy is designed to alter, at the hand of state regulators, the "just" and "reasonable" rates previously established pursuant to FERC-approved tariff or wholesale market design.

If allowances were given away without charge to upstream sources, which is where both RGGI and California may regulate, state credits could subsidize out-of-state businesses. If California did not auction allowances to emit carbon, it would instead divest without charge California carbon allowances to cover power generated by many out-of-state power generation sources. While approximately 20% to 25% of California's power is imported, approximately 55% to 60% of California's electricity-related GHG emissions that would need these allowances are associated with that smaller share of imported power. Therefore, a decision to distribute carbon allowances without charge to first-sellers of wholesale power would distribute up to 60% of carbon credits to cover out-of-state power or to external power generation businesses operating out-of-state.

States are hesitant to subsidize out-of-state business, which motivates auctioning these allowances. The natural incentive of any state is not to give away credits created by regulation in State A, which have a significant resale value, to entities in State B, which does not regulate carbon. States cannot directly regulate assets operating outside their borders, however, pursuant to U.S. law.

The articulated rationale in many of the states moving toward auction of carbon emission allowances is to prevent affected electric generators from earning excess or windfall profits from free carbon allocation and the operation of the wholesale power market. An explicit design objective of an auction is to raise the cost of high-carbon wholesale power production and reduce windfall or excess profits to any power generators.[12] Thus, the auction is designed to impose and regulate carbon costs by altering the market prices at which power from different generation sources trades at the wholesale level from wholesaler to retailer.

For example, carbon costs can have a significant impact on the ultimate price of electricity. If approximately one ton of carbon is created per megawatt hour (of electricity produced from coal-fired generation), then a cost of $10 per ton to purchase a carbon allowance or credit translates into an increase in cost of approximately $0.01/kWh in the price of electricity, for which the wholesale price is about $0.06/kWh at present.[13] At $10 per ton for the cost of CO_2 allowances or credits, this would add approximately $75 per year, or 10%, to the cost of power for the average household electricity bill of approximately $1,500 per year (more in certain high-cost states).

However, it is not even clear that $10 per ton for a carbon credit is realistic. European Union CO_2 emission credits have traded at twice this price, and as much as three times in extremes. In addition, to be able to stabilize CO_2 atmospheric concentrations at 650 ppm (the current levels are about 380 ppm), it is projected that a price of $20 per ton would be required for CO_2 credits. To stabilize CO_2 emissions at a lower level of 550 ppm, it is projected that credits would trade at $50 per ton. This cost would consume 1% to 2% of gross domestic product in developed countries, such as the United States.[14] The various projections of allowance trading costs in the United States are illustrated in figure 14-1.

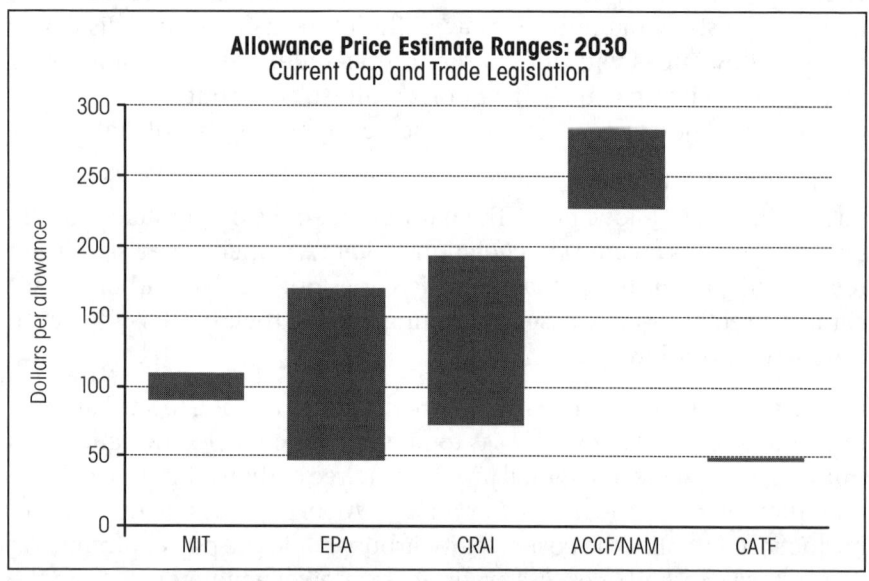

Fig. 14-1. Various projections of allowance trading costs in the United States *(Sources: Various studies cited, all prices adjusted for inflation to 2007 dollars)*

Carbon auction results

The RGGI first auction of emission allowances occurred in September 2008. The price of allowances auctioned was $3.07/ton; at subsequent auctions it has continued to trade below $3.50/ton. If a low RGGI allowance price turns out to be accurate, it might signify that CO_2 reductions are not in fact being accomplished because a surplus of credits and offsets are driving prices lower.

Operating costs affect which units are dispatched to run by the ISOs, the dispatch order of designated power generation, and which units are ordered not to run. This ISO regulatory mechanism does two things whose physical implications are important to understand. First, baseload plants designed to operate on a constant basis, which instead are pushed to the end of the dispatch queue, can become so uneconomic to run in this altered mode that they instead may be deactivated. Second, most of the high-carbon power plants are older baseload Brayton cycle plants designed to operate around the clock, not to cycle on occasionally to meet a peak need. If these baseload plants are reassigned because of cost of carbon regulation to a service mode where they are cycling on and off, they experience metal fatigue and metal creep, which under certain conditions can destroy the operating efficiency and capital equipment of their prime movers. So this change in dispatch order can become a death sentence for the operational, and in some cases physical, longevity of certain high-carbon generating assets banished to the end of the dispatch queue.

Various documents and reports issued by the regional RGGI Staff Working Group state that an expressed objective of the RGGI MOU is to modify the dispatch order and the carbon intensity of the existing portfolio of power generation units.[15] This could run afoul of the Federal Power Act's grant of exclusive federal jurisdiction over such wholesale matters and be preempted by the Supremacy Clause and the filed rate doctrine.

Here again, motive matters. As the Supreme Court articulated in *Pacific Gas & Electric Co. v. California Energy Resources Conservation and Development Commission,* the articulated motive of the agency regulating power resources will be taken at face value as the true motive for purposes of constitutional preemption analysis.[16]

The Bright Line: Federal Preemption of State Carbon Regulation

Sections 205 and 206 of the Federal Power Act empower the FERC to regulate rates for the interstate or wholesale sale and transmission of electricity. The act creates a "bright line" between state and federal jurisdiction, with wholesale power sales falling clearly and unequivocally on the federal side of the line.

FERC jurisdiction preempts state regulation of wholesale power transactions and prices. Where federal law occupies the field and there is evidence of a pervasive federal scheme in a given area, courts will find state or local legislation preempted.[17] Even where there is no evident congressional intent to federally occupy a field, the conflict principle requires that a court strike inconsistent state or local law.[18] State regulation is not allowed to veto the regulatory scheme of a superior level of government.[19]

Consider the delineation between federal and state authority under the Federal Power Act. The North American power grid is comprised of many individual pieces, owned by the local transmission companies, which operate under the overlapping jurisdiction of 55 state and Canadian provincial government agencies, as well as three national regulatory authorities (see chapter 12). Within the United States, FERC exclusively regulates every element of wholesale power sales. The Federal Power Act defines "sale at wholesale" as any sale to any person for resale.[20]

FERC's exclusive power is even broader than just wholesale power sales. FERC also regulates power generation (to a limited degree), power transmission in interstate commerce, and interstate power sales.[21] "FERC jurisdiction is plenary and extends to all sales in interstate commerce."[22] FERC does not regulate the local distribution of power, power solely in intrastate commerce, or the self-generation and use of power where there is no wholesale sale.[23] FERC jurisdiction can extend from the point of the power's origin on the basis that the entire sale affects interstate commerce.[24]

The quantity of power transacted is not a limiting factor on FERC jurisdiction. There is no statutorily or judicially imposed threshold amount of interstate sale of power that triggers FERC jurisdiction. Although the amount of power an electric utility may place in interstate commerce is de minimis compared to the same utility's sales in intrastate commerce, FERC may assert its regulatory authority over such a utility.[25] If a small amount of intrastate power is commingled with interstate power, the entire amount of power becomes "interstate" for purposes of vesting FERC with the authority to exercise jurisdiction.[26]

As the United States Court of Appeals for the Ninth Circuit has remarked, and the Supreme Court confirmed, when combined with federal preemption precedent, energy market regulatory reforms have contributed to "a massive shift in regulatory jurisdiction from the states to the FERC."[27] RGGI auctions are designed to affect the prices and terms of (1) wholesale power transactions, (2) including interstate power transactions, and (3) transmission of high-carbon power into the state. All three of these are subject to exclusive federal jurisdiction; state authority is preempted. In sum, there is limited prudential or other residual state authority of any kind over wholesale sales of power subject to exclusive federal jurisdiction after these most recent Supreme Court decisions. Therefore, their influence over wholesale terms and prices is preempted as a matter of constitutional law.

The Filed Rate Doctrine

If a utility or independent power producer is subject to FERC jurisdiction and regulation, state regulation of the same operational aspects is preempted as a matter of constitutional law.[28] Principles of preemption require a state regulatory agency to accept and pass through in retail rates all cost items deemed by FERC to be "just and reasonable," and which are otherwise allowed.[29] Therefore, a FERC determination regarding any aspect of a wholesale price is universally binding.

Supreme Court decisions

The so-called filed rate doctrine holds that state regulatory agencies may not second-guess or overrule on any grounds a wholesale rate determination made pursuant to federal jurisdiction.[30] The Supreme Court in 1988, 2003, and 2008 upheld the filed rate doctrine.[31] Five years ago, the Supreme Court clarified that there is little residual authority to reserve a state role in determining the price or terms of wholesale power transactions. This would mean that states may not retain residual authority to alter the wholesale market cost of high-carbon power with mandatory purchase of additional regulatory credits or assets. Such sale of emission rights, rather than free allocation of regulatory credits or assets designed to increase the trading price of certain power resources, contrary to FERC tariffs.

Any state's deliberate attempts to design RGGI regulations to tilt the wholesale electric market operation, power pricing, and dispatch order in wholesale markets (operating pursuant to FERC-approved tariffs) runs counter to the Court's 2003 *Entergy Louisiana, Inc. v. Louisiana Public Service Commission* decision. Where it is a wholesale or interstate power price matter, FERC alone has jurisdiction. This is exactly the point where certain state carbon regulation has declared that it would attempt to influence the ultimate cost of wholesale power.

Moreover, attempts by states indirectly or directly to promote higher wholesale energy prices for certain higher-cost low-carbon renewable energy projects have been stricken by the courts. In 1994, the United

States Court of Appeals for the Ninth Circuit rejected the California Public Utilities Commission's claim that it had independent authority to regulate the prices and terms for such low-carbon renewable power sales.[32] Promotion of certain types of low-carbon renewable fuels for power supply, via a price preference above and beyond the FERC-established price of other wholesale power transactions, was held preempted by the Federal Power Act and stricken.

Precedent holds that a higher price set by California for renewable low-carbon electric power supply sources is not permissible.[33] If a state is prohibited from inflating the quantity of certain renewable resources by setting higher wholesale prices for such favored technologies, it follows that it could also be prohibited from accomplishing the same "tilt" in wholesale prices by the opposite mechanism: inflating the wholesale operating costs of less desirable high carbon-emitting generation resources. If it mandates that high-carbon generating resources purchase more carbon allowances, it achieves the same relative tilt in favor of certain resources.

In each of the RGGI states, there is no approval by FERC to run plants on anything other than the market bid rules that are approved and in place for purposes of ISO operation. Those rules do not countenance a state imposing a differential auction obligation designed and announced by the state to raise the price of particular (high-carbon) wholesale generation within the state. A state law may not frustrate the operation of federal law, even if the state legislature has valid purposes for the legislation.[34]

State law is not allowed to preempt federal determinations by layering on additional requirements not contained in federal law.[35] The wholesale price determination is reserved exclusively to federal authority.[36] The filed rate doctrine extends to non-rate matters as well. The Federal Power Act precludes all state regulation of interstate wholesale power transactions.

Affirmative federal monitoring obligation over state actions

During the 2000-2001 California power shortage, the crisis was significantly linked to California's allegedly flawed restructured retail power market design and regulation. When prices subsequently fell, California attempted to be excused from the very wholesale power supply contracts that it had leveraged into place on reluctant wholesale power suppliers.[39] The state's legal argument was that the wholesale power contracts were the exclusive province of federal jurisdiction by FERC, and FERC had not sufficiently policed the wholesale market. This event occurred before the Supreme Court decision in *Entergy*.

A majority of the cases before the United States Court of Appeals for the Ninth Circuit affirmed this theory, which was later affirmed by the U.S. Supreme Court. In *Public Utility District No. 1 of Snohomish County, Washington v. FERC,* FERC, as the wholesale power regulator, must protect the state (and other stakeholders) against the state's own regulatory actions or mischief. FERC not only has exclusive authority unaffected by any state actions over wholesale power markets, but FERC has an ongoing obligation to continually monitor and police these markets against state interference.

Auction of allowances significantly impacts the price and terms of wholesale power transactions, which in turn alters the dispatch order of wholesale power generation within the state and ISO power pools. The RGGI states and California operate under FERC-approved terms and conditions for ISOs. All power sold into the grid, which is managed by the ISO, is sold under wholesale terms and conditions that are part of its approved FERC tariff. These tariffs do not contemplate state efforts to tilt the wholesale market because of carbon policies. The first challenge to the New York RGGI carbon regulation was brought by Indeck Energy in 2009, alleging various legal violations of the entire RGGI system in New York, which is an opening shot of legal challenges. New York settled this suit, granting the plaintiffs everything they sought rather than allow this case to go to decision and create precedent.

Reserved State Legal Discretion

From any type of source, moving electrons are power. There is no engineering difference in the electricity end product. It is clear that the state can regulate non-price aspects of the power sale market within state boundaries.[40] Within this general authority, states have regulated what electric facilities can be sited, where they can be sited, environmental standards of plant operation, and the mix of demand-side and supply-side resources. For example, California by statute prevents the construction of new nuclear power facilities until such time as there is a solution to the long-term nuclear waste disposal problem.[41]

If properly utilized, a state could manage the carbon intensity of its power supply not by controlling its borders (see chapter 13), but by specifying eligible environmental parameters for regulated purchasers of power by regulated utilities. There is mixed jurisprudence on how far a state can go. As one example, a New York decision held that a state cannot compel a utility to purchase power from a particular wholesale source.[42]

Other courts have allowed states to regulate the mix of generating/efficiency resources that regulated utilities must procure.

> [U]nder state authority, a state may choose to require a utility to construct generation capacity of a preferred technology or to purchase power from the supplier of a particular type of resource. The recovery of costs of utility-constructed generation would be regulated by the state. The rates for wholesale sales would be regulation by this [FERC] Commission on a cost-of-service or market-based rate basis, as appropriate.[43]

What a state cannot do is to attempt to influence the price of a wholesale power transaction, which is exclusively within FERC jurisdiction:

> "We [FERC] cannot ascertain at this date any legal basis under which states have independent authority to prescribe rates for sales by QFs [qualifying facilities] at wholesale [to utilities] that exceed the avoided cost cap contained in PURPA."[44]

Moreover, there are constitutional limitations to a state directly taking regulatory action that affects persons or property in another state, which limit extraterritorial efforts of any state to regulate power emanating from a different state.[45] Auction of allowances by a state can violate the Supremacy Clause constitutional limitation on states not affecting either directly or indirectly the wholesale terms and prices of transactions in power. The filed rate doctrine provides strict limits on the state role in the power sector, regardless of motives.

To conclude this third section of this book, the next chapter turns to the new legal requirement of *additionality* being imposed in most carbon regulatory schemes. It also is an element of the Waxman-Markey carbon regulation bill at the national level. Additionality is designed to condition the novelty of carbon offsets. It works as a limitation on the currency of carbon. Yet, there is little consensus on additionality. Its pros and cons are next explored.

Notes

1 Weinzimer, Lisa. 2008. Schwarzenegger Advisor Says States, Regions Will Take Lead on Climate Program. *Electric Utility Week*. June 16: 7.

2 For a discussion of various power auction practices and theories, see Ferrey, Steven. 2009. *The Law of Independent Power*. Eagan, Minnesota: Thomson/West Publishing: chap. 9.

3 For a discussion of the operation of some of these ISO regional energy control areas, see ibid., sec. 10.87.

4 New York State Department of Environmental Conservation. 2008. *Revised Regulatory Impact Statement 6 NYCRR Part 242, CO_2 Budget Training Program*. 43. http://www.dec.ny.gov/docs/air_pdf/242ris.pdf.

5 ISO New England, Inc. 2008. *ISO New England Manual for Market Operations: Manual M-11*. http://www.iso-ne.com/rules_proceds/isone_mnls/.

6 See, New York State Department of Environmental Conservation. 2007. *AES & Dynegy, Comments on Proposed 6 NYCRR Part 242, CO_2 Budget Trading Program, Revisions to 6 NYCRR Part 200*. December 24: 101.

7 Wood, Lisa. 2008. Connecticut's AG Challenges Governor's Push to Channel Some RGGI Money to Ratepayers. *Electric Utility Week*. July 21: 8.

8 California Public Utilities Commission and California Energy Commission. 2008. *Joint Staff Paper on Options for Allocation of Greenhouse Gas Allowances in the Electricity Sector*. R. 06-04-009, Dec. 07-OIIP-01.

9 California Energy Commission. 2007. *Pre-Workshop Comments of Dynegy on Allocation Issues*: 8. R.06-04-009, D.07-07-018, Cal. P.U.C. LEXIS 330. http://docs.cpuc.ca.gov/efile/CM/75023.pdf.

10 Grenfell, Kristin. Natural Resources Defense Counsel. 2008. *Memorandum to California Air Resources Board Staff RE: NRDC Comments on Allowance Allocation Issues*. http://www.arb.ca.gov/cc/scopingplan/pgmdesign-sp/meetings/031708/NRDC_comments_from_031708_Program_Design_Technical_Stakeholder_Workgroup_Meeting.pdf.

11 Because the value of the allowances will be included as a cost in the generators' bids to supply electricity, the price of electricity will be the same whether the allowances are given away at no cost to generators or generators must purchase allowances. An allowance giveaway, therefore, means generators are able to substantially increase their revenues (and, hence, profits) under a program like RGGI because they pass on the cost of a commodity they obtained at no charge. This has been referred to as "excess revenues," and these excess revenues occur at the expense of electricity consumers. New York State Department of Environmental Conservation. 2006. *Notice of Pre-Proposal of New York RGGI Rule*. http://www.dec.ny.gov/regulations/26450.html.

12 New York Compilation of Codes Rules and Regulations. Title 6, sec. 242.

13 For natural gas-fired generation, the CO_2 emissions are approximately half this coal-related amount, with approximately one-half ton of CO_2 emissions per megawatt hour of electricity generated. Marion, John et al. *Controlling Power Plant CO_2 Emissions: A Long-Range View*. National Energy Technology Laboratory: 3, fig. 3a. http://www.netl.doe.gov/publications/proceedings/01/carbon_seq/1b2.pdf. This source depicts CO_2 emissions from electricity and heat generation sectors in Annex II countries.

14 See, Bureau of Economic Analysis. 2008. *National Economic Accounts.* http://www.bea.gov/national/index.htm#gdp. This source shows the current dollar and gross GDP.

15 New York State Department of Environmental Conservation. 2006. *Notice of Pre-Proposal of New York RGGI Rule.* http://www.dec.ny.gov/regulations/26450.html.

16 *Pacific Gas and Electric Co. v. State Energy Resources Conservation and Development Commission.* 461 U.S. 190, 216 (1983).

17 See, *City of Burbank v. Lockheed Air Terminal, Inc.*, 411 U.S. 624 (1973). This case held that the federal government occupied field of noise regulation for aircraft.

18 See, *Rice v. Santa Fe Elevator Corp.*, 331 U.S. 218, 230 (1947); *Hill v. Florida ex rel. Watson*, 325 U.S. 538, 541–42 (1945); *Rombom v. United Air Lines, Inc.*, 867 F.Supp. 214 (S.D.N.Y. 1994). This case held that state law cannot stand as an obstacle to accomplishment and execution of the full purposes and objectives of Congress.

19 See, *Granite Rock Co. v. California Coastal Commission*, 480 U.S. 572 (1987), reversing 768 F.2d 1077 (9th Cir. 1985).

20 16 U.S.C. sec. 824(d).

21 Ibid., sec. 824(a)–(b).

22 *Northern States Power Co. v. Minnesota Public Utilities Commission*, 344 N.W.2d 374, 378 (Minn. 1984), *cert. denied*, 467 U.S. 1256 (1984).

23 See, *Connecticut Light & Power Co. v. Federal Power Commission*, 324 U.S. 515 (1945); *City of Batavia v. Federal Energy Regulatory Commission*, 672 F.2d 64, 68 n.2 (1982). FERC regulates wholesale transactions; states regulate retail transactions.

24 *Jersey Central Power & Light Co. v. Federal Power Commission*, 319 U.S. 61, 70–72 (1943).

25 *Federal Power Commission v. Southern California Edison Co.*, 376 U.S. 205 (1964); *Arkansas Power & Light Co. v. Federal Power Commission*, 368 F.2d 376 (8th Cir. 1966).

26 *United States v. Public Utilities Commission*, 345 U.S. 295 (1953); *Cincinnati Gas & Electric Co. v. Federal Power Commission*, 376 F.2d 506 (6th Cir. 1967), *cert. denied*, 389 U.S. 842 (1967); *Public Service Co. of Indiana v. Federal Power Commission*, 375 F.2d 100, 104 (7th Cir. 1967).

27 *Snohomish County Public Utility District No. 1 v. FERC*, 471 F.3d 1053, 1066, (2006) *affirmed in part and reversed in part sub nom. Morgan Stanley Capital Group, Inc. v. Public Utility District No. 1*, 128 S. Ct. 2733 (2008); *Entergy Louisiana, Inc., v. Louisiana Public Service Commission*, 539 U.S. 39 (2003).

28 *Arkansas Power & Light Co. v. Federal Power Commission; Nantahala Power & Light Co. v. Thornburg*, 476 U.S. 953 (1986); *New England Power Co. v. New Hampshire*, 455 U.S. 331 (1982).

29 *In re* Appeal of Sinclair Machine Products, Inc., 498 A.2d 696, 706 (N.H. 1985).

30 However, the Supreme Court has determined that Congress, in enacting the Federal Power Act, intended to vest exclusive jurisdiction in FERC to regulate interstate wholesale utility rates. *Federal Power Commission v. Southern California Edison Co.*

31 See, *Nantahala Power & Light Co. v. Thornburg.* "This Court has held that the filed rate doctrine applies not only to the federal-court review at issue in Montana-Dakota, but also to decisions of state courts."; *Mississippi Power & Light Co. v. Mississippi ex rel. Moore*, 487 U.S. 354, 372 (1988). This case held that the filed rate doctrine applies without exception to state regulation of interstate holding companies; *Entergy Louisiana, Inc., v. Louisiana Public Service Commission.* This case held that no residual prudency power of the states can alter a federal rate or term.

32 *Independent Energy Producers Association v. California Public Utilities Commission*, 36 F.3d 848 (9th Cir. 1994). This case found no separate basis for the state PUC to act to establish a premium price for renewable low-carbon power projects.

33 Southern California Edison Co. 70 FERC par. 61,215 (1995). This hearing held that the costs of renewable energy are not to exceed the market or bid price of all other sources of energy makes ratepayers indifferent as to the procurement of wholesale power. *Requests for reconsideration denied*, 71 FERC par. 61,269 (1995).

34 See, e.g., *Perez v. Campbell*, 402 U.S. 637 (1971). *Superseded by statute*, Act of Nov. 6, 1978, Public Law 95-598, 92 Stat. 2593.

35 See, e.g., *Granite Rock Co. v. California Coastal Commission*, 768 F.2d 1077, 1083 (9th Cir. 1985), reversed on other grounds, 480 U.S. 572 (1987).

36 *Entergy Louisiana, Inc., v. Louisiana Public Service Commission.*

37 *Northern Natural Gas Co. v. State Corporation Commission*, 372 U.S. 84, 90-91 (1963); *accord* Nantahala Power & Light Co. v. Thornburg, 476 U.S. 953, 966-67 (1986).

38 *Nantahala Power & Light v. Thornburg*, 476 U.S. 953; *Mississippi Power & Light Co. v. Mississippi ex rel. Moore*, 487 U.S. 354, 371 (1988); *accord Mississippi Industries v. Federal Energy Regulatory Commission*, 808 F.2d 1525, 1535-49 (D.C. Cir. 1985), *cert. denied*, 484 U.S. 985 (1985).

39 *Snohomish County Public Utility District No. 1 v. FERC*, 471 F.3d 1053, 1067.

40 See, S. Cal. Edison Co., 70 FERC par. 61,215, par. 61,676-77 (1995).

41 *Pacific Gas & Electric Co. v. State Energy Resources Conservation and Development Commission*, 461 U.S. 190 (1983) (upholding the Warren-Alquist Act).

42 *Consol. Edison Co. v. Public Service Commission*, 472 N.E.2d 981, 987 (N.Y. 1984).

43 Southern California Edison Co., 70 FERC par. 61,215, par. 61,676 (1995).

44 Ibid.

45 See, *Shaffer v. Heitner*, 433 U.S. 186, 197 (1977). "[A]ny attempt 'directly' to assert extraterritorial jurisdiction over persons or property would offend sister States and exceed the inherent limits of the State's power."; *C. & A. Carbone, Inc. v. Town of Clarkstown*, 511 U.S. 383, 392–93 (1994); *Cf. Baldwin v. G.A.F. Seelig, Inc.*, 294 U.S. 511 (1935). This case invalidated New York prohibition on cheaper out-of-state milk being sold in state for less than New York minimum price, thus discriminating against the competitive advantage of foreign producers.

15) LEGAL ADDITIONALITY REQUIREMENTS FOR CARBON OFFSETS

What Is Additionality?

The greatest concerns about carbon trading among stakeholders are the requirements of *additionality* and verification of offsets.[1] Additionality creates a major legal and regulatory disconnect in these carbon laws, especially regarding renewable power options. Additionality is the requirement in most of these carbon control statutes or regulations that only *additional* or non-business-as-usual carbon-reduction projects legally qualify to create carbon *offsets*; offsets create tradable credits for compliance with these carbon policies.[2] Such offset credits, which are embodied in all international and U.S. state carbon laws enacted and proposed to date, can be earned and traded among regulated industries, such as power generators, for compliance with the carbon laws. They control the common currency of carbon.

However, some of these carbon programs have specifically excluded all renewable energy projects from being deemed "additional" or eligible to create this new carbon currency (see chapter 16). The legal rationale is that renewable power is abundantly promoted by a host of other legal incentives, such as tax credits, accelerated tax depreciation, creation of renewable energy credits (RECs), or system benefit charges to promote renewable power (see chapter 18). These are deemed to be constructed anyway and are not additional or justified due solely to a carbon control program. To allow renewable projects to legally "double-dip," as both renewable projects and carbon reduction projects, even though they truly serve dual purposes technically, is defined as failing additionality." The first carbon reduction program in the United States, the Regional Greenhouse Gas Initiative (RGGI), commenced in January 2009 in 10 northeastern states (see chapter 7), takes this legal position by barring renewable energy projects as never "additional," as set forth in the following paragraphs.[3]

The quid pro quo for offsets has been the requirement for additionality: additionality has become the meta-legal requirement of qualifying a carbon emission offset under the international Kyoto Protocol and the European Union Emission Trading Scheme (EU-ETS), as well under RGGI. Additionality even legally trumps any requirement to have an offset substitute renewable energy generation for high-carbon-emitting conventional power generation. Under the only offset results available to date, the Clean Development Mechanism (CDM) Kyoto offsets are principally avoiding renewable energy projects in favor of higher-return abatement of hydrofluorocarbons (HFCs) and burning of methane without using it to generate power (see chapter 10).

So where did the legal definition of additionality emanate from and how does it influence or contort U.S. and world carbon policy? All certified emission reductions (CERs) under the Kyoto Protocol CDM are required by the protocol to be voluntary, real, and additional to any that would occur in the absence of the CDM credit system.[4] There are at least eight different tests of additionality, none of which are commonly accepted as credible. Retailers of offsets often could provide little information or claimed that their offsets were additional, but the U.S. GAO found that some sellers could not explain how they defined "additional" and provided little verifiable information to buyers.[5] This made offsets less credible and could compromise the integrity of a carbon reduction system.

At its core, additionality is a legal, rather than technical or engineering, concept. Additionality has no technical analogue, as every ton reduction of carbon has scientific value in reducing atmospheric concentrations, regardless of whether additional or not or how or why obtained. Stakeholders surveyed by the U.S. GAO believed that additionality is not a critical factor and barriers to entry of carbon reduction strategies must be lowered, especially since there is no accepted definition of it. The actual impact of additionality is to reduce the net quantity of available traded allowances in a regulatory system. The additionality requirement does nothing to reduce total GHG emissions, but it does reduce the availability of eligible GHG reduction offset credits. This reduction in supply will tend to increase the market clearing price in trading of eligible GHG reduction offsets.

The History of Additionality

Additionality as a legal concept originally was added to the three other requirements for U.S. Clean Air Act Emission Reduction Credits (ERCs) for nitrous oxide (NO_x), and later was adapted by the RGGI states for carbon offsets. There have been five total prior cap-and-trade emission markets established in the United States: acid rain (SO_2), NO_x summer ozone budget program (12 state NO_x), Clean Air Interstate Rule (CAIR) (NO_x and SO_2), the Mercury Rule, and Regional Clean Air Incentives Market (RECLAIM) in Southern California (NO_x and SO_2). All but RECLAIM have been administered by the U.S. EPA. All allowances under each of these five prior programs were not auctioned but were allocated free to traditional emitters of the pollutant based on average unit heat input (e.g., acid rain program) or depending on state-specific programs (e.g., summer ozone).

But all has not gone smoothly with these prior cap-and-trade emission programs. The two most recent of these cap-and-trade programs, the CAIR NO_x and SO_2 trading rules[6] and the mercury trading rules,[7] were ruled legally impermissible and stricken in 2008 by the United States Court of Appeals for the District of Columbia Circuit.[8] Most recently, in mid-2008, the court of appeals struck and remanded EPA's CAIR, which required 28 states and the District of Columbia to reduce 61% of regional NO_x emissions and 73% of regional SO_2 emissions below 2003 levels by 2015 starting in 2009, and eliminate "significant" contributions to downwind states' air pollution.[9] The court declared that it found "more than several fatal flaws in the rule" and remanded the matter to EPA to promulgate a new rule consistent with the opinion. This has raised questions as to whether EPA can use a cap-and-trade system to address National Ambient Air Quality Standards (NAAQS) attainment in downwind states under the Clean Air Act for fine particulate matter, SO_2, NO_x, and ozone.

The price of SO_2 allowances traded plunged by half after the CAIR decision; the price of allowances was less than the cost of lime to operate the scrubbers that actually reduce power plant SO_2. During the 2000-2001 California electricity crisis when additional high-emissions resources were required, the price of RECLAIM allowances skyrocketed and state pressure to produce more electricity caused an exceedence of the NO_x cap and an eventual removal of electric generation from RECLAIM coverage during the crisis.[10]

Monitoring and Verification of Additional Carbon Controls

Additionality has become the regulatory test for integrity of offsets, although there is no certain or accepted tool to accomplish this determination of what is or is not additional. Regulatory additionality is made more complex because it requires a baseline against which to measure reductions: reductions are only qualified as additional when above a project-specific or standardized baseline. The baseline requires accurate measurement, correct counting for attribution, and permanence of the reduction. Double-counting can occur, for example, where one region or nation installs a renewable power generation project, another region or nation purchases the power from the project, and yet another nation or state purchases the virtual renewable energy credit(s) associated with the project. Which can claim the carbon reduction attributes?

This becomes even more challenging where one region regulates carbon and another does not. Accessing possible leakage into the system of regulated carbon emissions from other unregulated sources poses an additional problem (see chapter 13). Because power moves almost at the speed of light and is not precisely traceable as to source, and offset credits as a virtual regulatory creation move subject to differentiated legal rules, there can be conflicts and challenges. However, additionality is the most significant challenge.

The measures of additionality for purposes of environmental emission credits trading are extremely imprecise. First, developing a baseline "business-as-usual" scenario, against which to measure what is additional, is subject to variable assumptions. Against this business-as-usual-scenario, actual carbon savings are calculated and assumed to be an emission reduction. Many additional NO_x ERC-creating projects traditionally earn credits by shutting down for economic or political reasons, rather than actually implementing emission-saving measures associated with continued operation or output. Similar reductions or shutdowns may or may not also be eligible for creating carbon offsets. The GAO concluded that the verification of additionality was burdensome and overly bureaucratic, extremely subjective and often manipulated, and frustrated stakeholders.[11] It is currently controversial everywhere.

Second, looking at international precedent, calculation of Kyoto Protocol CDM project-based offsets are generally linear extrapolations of models, which may not be representative of what actually occurs on the ground, as shown by the significant overestimation of CDM credits.

For the first 175 CDM projects that issued CERs, the validation procedure overestimated the number of CERs produced by approximately 27% on average, with a standard deviation error of 42.5%.[12] Verification can be an inside game: three firms have been collectively involved in verifying more than 80% of the first 740 CDM projects that were registered under the Kyoto Protocol.

Third, issuing credits on a project-by-project basis increases inaccuracy because any pattern of deviation from actual realized values is then multiplied over a large number of small transactions. The fact that national compliance was shown so easily in 2006 in the European Union carbon trading scheme, after early predictions of shortfall, indicates that flexibility to demonstrate compliance on paper often masks the true reduction. Despite this apparent compliance with goals to significantly cut annual GHG emissions, rather than declining in either 2006 or 2007, world GHGs actually climbed 2.9% from 2006 to 2007 and increased in the European Union.

In a market with tradable credits, additionality does not reward the obvious or best societal investments, precisely because such investments are economically doable or compelled anyway, and therefore not additional solely because of the carbon control program. In other words, to satisfy additionality, energy investors invest in something that is marginal to qualify it as additional to normal investments. Additionality requirements for offsets increase transaction costs for certifying credits. Transaction costs range up to 20% of trading prices for the credit.

Environmental groups have questioned the additionality of renewable energy projects, if their construction cannot be proven to be because of the value of the offset sale. Renewable power investments are not recognized as carbon offsets because "... the emission reduction doesn't occur at the site of the renewable generator," but in backing out other carbon-intensive generation.[13] Figure 15-1 illustrates the renewable composition by type of project: wind dominates. Under the current RGGI construct, the coal, power, and railroad industries have threatened some states with suit over the RGGI program.

It is necessary to distinguish offsets and allowances as to additionality requirements. While RGGI includes the *offset* requirement of additionality, there is no concept of additionality in either the RGGI original allocation of carbon *allowances* among the 10 individual RGGI states, or through those states to those who acquire such allowances by either auction or allocation.

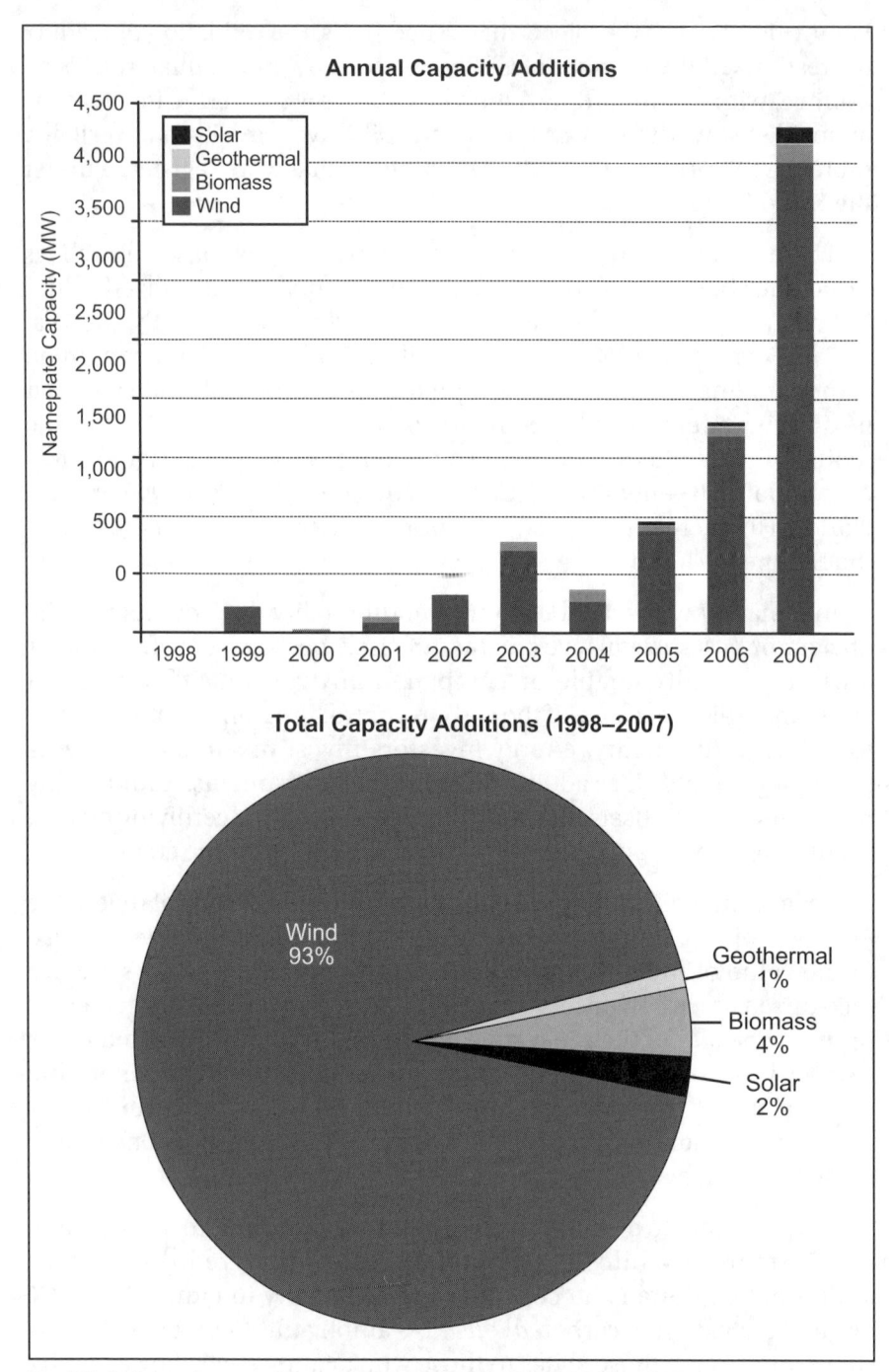

Fig. 15–1. Annual renewable power capacity additions, 1998–2007

What Qualifies as Additional?

Vintage allowances over time

In the RGGI scheme, there is an additionality requirement for early reduction RGGI *allowances* achieved at fossil fuel projects.[14] This transforms such early allowances, which otherwise typically do not require additionality, to the equivalent type of additionality requirement for carbon offsets which also require additionality. Ordinary allowances are merely purchased through the auction and trading system regardless of additionality. Early compliance credits also were incorporated in the U.S. credit-trading program for SO_2, which did not require additionality be demonstrated for its early compliance.[15] To make things even more disconsonant, by definition, RGGI offsets cannot be created at fossil-fuel-fired power projects, although early-reduction allowances will be created at existing fossil-fuel-fired projects and require parallel additionality as offsets.[16]

If the objective of RGGI is to hold CO_2 annual emissions constant at their historic levels, ignore past emissions, and then permanently reduce CO_2 after 2015, allowing 2006–2008 early reductions to count and be shifted forward in time to be used as if they were later-period reductions does not directly contribute to this. However, if continued long-term, there may be value in allowing credit for early reductions. Since CO_2 emissions become a carbon concentration in the atmosphere lasting for a century, any additional reduction accomplished in the past century is of value globally, and might be of creditable value in a regulatory system.

Some RGGI states have gone even further, qualifying early allowances not requiring additionality. Massachusetts, which prior to joining RGGI regulated CO_2 emissions from six large fossil-fueled power generation facilities in the state, also allows some of these terminated prior program compliance actions to create transferrable credits qualifying as RGGI offsets.[17] This will apply to otherwise RGGI-ineligible offset projects required by other law and involves state set aside from Massachusetts' RGGI allowance portfolio. In essence, it allows prior expenditures on carbon control projects required by law in Massachusetts under a different regulation to create RGGI allowances from early efforts, which by definition certainly are not additional to what was required by law, and therefore normally ineligible.

Offsets

Additionality is required of all RGGI offsets. Renewable power is not required of any offsets. In fact, renewable energy projects are not allowed to create RGGI offsets.[18] Methane is being flared at certain offset projects to garner credits, even though such flaring is not additional, and often does not generate power but consumes it as a waste gas.

While RGGI, like Kyoto, requires offset additionality and does not require renewability, RGGI goes even further to establish bars so that renewable power projects expressly do not qualify to create offsets. The media has questioned the credibility of carbon offsets and the efficacy of such offsets. "The vast majority of offsets are, at some level, just rip-offs," according to a former Clinton administration official.[19]

California is considering allowing RECs (see chapter 18) to also count for carbon reduction. This is the opposite of what RGGI allows. Some activists are trying to limit out-of-state offset credit for out-of-California renewable energy project offsets. Southern California utilities are urging no restriction on out-of-state renewable energy credits. The California carbon program has not yet made its final decisions for additionality and renewability of offset projects, prior to its commencement in 2012. Neither have the western states nor midwestern states carbon programs. Limiting import of credits, when they are tied to production of electricity in interstate commerce, could raise constitutional Commerce Clause issues (see chapter 13).

Kyoto Conundrum

Additionality is in the text of the Kyoto Protocol: emission credits/offsets must be "supplemental to domestic actions for the purpose of meeting quantified emission limitations and reduction commitments."[20] With Kyoto, verifying additionality includes no requirement for more renewable resource deployment. Kyoto has additionality without renewability. There is no mandatory environmental or sustainability assessment in Kyoto projects or public input, which was rejected by the Kyoto developing countries as an infringement on host country sovereignty.

The impact of Kyoto CDM projects has not been to emphasize appropriate renewable investments in developing countries, but rather has served to create salable transferable additional credits for Annex I countries (see chapter 10). Unlike fossil fuels, renewable resources are widely disseminated across the globe. Every nation has significant renewable energy in some form—hydropower, sunlight, wind, agricultural biomass waste, wood, ocean wave power, etc.

But unless the post-Kyoto architecture segregates and promotes these technologies, developing nations will not deploy them sufficiently, instead opting to burn coal and other traditional fossil fuels. For instance, developers of Kyoto CDM projects in developing nations are trapping methane and flaring it, without turning it into free electricity in the process (see chapter 13). These easy solutions reduce GHGs, but perpetuate the need for electricity for the host community from other sources. Renewable energy projects account for 28% of CDM CERs; methane capture and flaring projects producing no electricity, mostly located at large landfills, coal mines, and animal feeding operations, account for 19% of CERs. Therefore, while the Kyoto Protocol CDM process encourages carbon reduction in developing countries, it does not always result in a substitution of renewable power for conventional fossil fuel power.

The Congressional GAO report in November 2008 stated that the "evidence indicates that the CDM has had a limited effect on sustainable development."[21] GAO concluded that by encouraging the lowest-cost means for a developer to reduce carbon, the CDM scheme disadvantages measures that contribute to sustainable development. It also concluded that CDM has not been successful in promoting technology transfer, and that such technology transfer was most likely to occur to assist renewable power development. GAO found that the emphasis on reduction of HFC-23 GHGs "do little to promote efficient energy use or contribute to long-term sustainable development objectives." It further concluded that developing countries that host CDM projects dilute the stringent contribution of CDM programs to sustainable development, because of competition for these projects to be located in their countries.

The world stands at a crossroads in time because in the next decade, there will be a massive investment in electrification of developing nations. According to Rajendra Pachauri, Kyoto IPCC chairman, "What we do in the next two to three years will determine our future." Once installed, those generating facilities will remain in place, contributing to global warming, or not, often for 40 years or longer. These choices in energy technology made now certainly will be the signature of the world carbon footprint for the remainder of this century, during which we may pass the point of no return in terms of global warming.

But under the current Kyoto Protocol, proper incentives are not present: "The CDM has, for a variety of reasons, been largely unsuccessful in encouraging real and significant changes in developing countries," according to the June 2008 report of an independent task force of the Council on Foreign Relations.[22] It found that 20% of the CDM projects would have occurred notwithstanding CDM qualification, and another study found that one-third of projects in India failed to demonstrate their additionality from what would have been otherwise implemented. It concluded that CDM "has been disappointingly ineffective at achieving its goal of effecting fundamental shifts toward cleaner energy production."

Additional Thoughts

The new regulatory/legal requirement for additionality is the least discussed, yet critically important, component of the new carbon control regulatory constructs. The three carbon regulation schemes in effect today (RGGI in the United States, EU-ETS, and the Kyoto Protocol) all require additionality of offsets. Additionality has emerged as the dominant and controlling meta-screen for legal qualification of offset credits in carbon regulation in the United States and around the world. It has become even more important than the supposed goal of substitution of renewable low-carbon power generation for traditional generation. As such, additionality has even worked an absolute prohibition against any renewable power in the new United States carbon regulatory schemes up until now.

Even the transition between state level and federal carbon regulation in the United States poses issues. Pending federal legislation creates an interesting conversion: the Waxman-Markey carbon legislation provides that any allowances issued before 2012 could be exchanged for federal allowances based on the average auction price for allowances issued in a given year.[23] Therefore, conversion of allowances or offsets issued or certified during the first three years of RGGI, or early reduction credits issued in California, can be converted to any new federal carbon currency.

The Kyoto Protocol allows renewable energy project CDM offsets, where the U.S. RGGI program specifically excludes renewable energy projects offsets. But even under the Kyoto Protocol where renewable CDM offsets are allowed, there is no widespread renewable energy technology deployment, in fact. Rather, the clear majority of CDM offset projects around the world are concentrated in one country and avoid renewable energy projects. Long term, these failures to facilitate the necessary transition to renewable energy in various countries threaten to be the critical shortcoming of the CDM program.

Additionality employs a new math where the necessary long-term investment does not add up. It is a key regulatory concept that must be retooled to certify offset projects that recognize and count technologies that shift the power-generating base to a more substantial renewable power component. It is a concept totally of regulatory math, and that math can be changed to comport with the renewable power necessity. It now works a disconnect between means and ends: the end being a long-term renewable power base and preservation of forest cover (see chapter 20), but the means employing additionality divert the means from the end.

This section of the book has highlighted several key legal and regulatory issues confronting all carbon control systems in the world. Some involve policy choices, while others involve legal obstacles and impediments. The next and final section of the book turns to the policy and regulatory interface between carbon control and renewable power options. We will examine the key renewable power regulatory mechanisms in Europe and the United States as well as the successful renewable power design for developing countries, and the role of natural renewable conversion of CO_2 in the world's forests. These can all be part of the bridge to a carbon-constrained future.

Notes

1. U.S. Government Accountability Office. 2008. *Carbon Offsets: The U.S. Voluntary Market Is Growing, but Quality Assurances Poses Challenges for Market Participants.* GAO-08-1048. (August): 25.

2. See Regional Greenhouse Gas Initiative. 2007. *Model Rule.* http://www.rggi.org/docs/model_rule_corrected_1_5_07.pdf; U.S. Congressional Research Service. 2008. Climate Change and International Deforestation: Legislative Analysis. *CRS Report for Congress.* RL34634. August 22: CRS-5, table 1.

3. Regional Greenhouse Gas Initiative, *Model Rule,* sec. 5.3(a)–(b).

4. Kyoto Protocol to the United Nations Framework Convention on Climate Change. 1998. Art. 12, sec. 5.

5. U.S. Government Accountability Office, *Carbon Offsets,* 30.

6. 70 Fed. Reg. 25,162. May 12, 2005.

7. 70 Fed. Reg. 28,606. May 18, 2005.

8. See, *New Jersey v. EPA*, 517 F.3d 574 (D.C. Cir. 2008). This case struck the mercury rule.

9. *North Carolina v. EPA*, Case No. 05-1244 (D.C. Cir. 2008).

10. For a discussion of the California electricity crisis, see Ferrey, Steven. 2004. Soft Paths, Hard Choices: Environmental Lessons in the Aftermath of California's Electric Deregulation Debacle. V*irginia Environmental Law Journal*. 23 (2): 251.

11. U.S. GAO. 2008. *International Climate Change Programs: Lessons Learned from the European Union's Emissions Trading Scheme and the Kyoto Protocol's Clean Development Mechanism*. GAO-09-151. (November): 46–48.

12. Hart, Craig. 2007. The Clean Development Mechanism: Considerations for Investors and Policymakers. *Sustainable Development Law & Polic*y. American University, Washington College of Law. 54 (Spring): 42. (utilizing UNEP data.)

13. Carbon Control News. 2008. Environmentalists Split over Support of Offsets for Plant Closures. *Carboncontrolnews.com*. August 25: 1, 19.

14. Regional Greenhouse Gas Initiative (RGGI), *Model Rule*, sec. 5.3(c).

15. Code of Federal Regulations. 40 C.F.R. 73.71(a)-(f).

16. RGGI, *Model Rule*, sec. 10.3(d)(2).

17. 310 Code of Massachusetts Regulations. 7.29.

18. RGGI, *Model Rule*, sec. 10.3(d)(2).

19. Carbon Control News. 2008. Rethinking Offsets. *carboncontrolnews.com*. Oct. 6.

20. Kyoto Protocol, art. 17.

21. U.S. GAO. 2008. *International Climate Change Programs: Lessons Learned from the European Union's Emissions Trading Scheme and the Kyoto Protocol's Clean Development Mechanism*. GAO-09-151. (November): 43.

22. Council on Foreign Relations. Independent Task Force. 2008. *Confronting Climate Change: A Strategy for U.S. Foreign Policy*. June: 21.

23. Cordner, Christine and Gail Roberts. 2009. RGGI Expects Compensation for Early Action if Federal Legislation Is Approved. *Electric Utility Week*. April 20: 23.

PART IV

CARBON REGULATION INTERFACING WITH RENEWABLE POWER:
RENEWABLE TOOLS FROM THE TOOLBOX

16) OFFSETTING CARBON: CREATING CREDITS FROM RENEWABLE POWER AND CONSERVATION

Given the requirement for *additionality* in offsets examined in the prior chapter, do world carbon control efforts lead to the actual physical addition of renewable energy projects or energy conservation? Renewable power and greater energy efficiency are a critical link in the effort to constrain carbon emissions in the 21st century. To a significant degree, these aspects are linked: long-term carbon reduction depends on shifting the power generating base more to renewable power production that does not emit net carbon emissions, and using power resources more efficiently. Renewable technologies have been abundant for decades; carbon regulation was agreed to in principal 15 years ago by world nations, and carbon regulation in the European Union has been in place since 2005. Are there positive indicators of successful substitution of renewable energy or energy conservation?

This logical path has not yet been evident. In the United States, the Obama energy legislative initiatives have stressed both carbon control and renewable energy development. However, at the state level, RGGI carbon regulation in the United States does not allow renewable energy projects (other than methane destruction) to create tradable value as carbon offsets, and the degree of eligibility of energy efficiency remains still to be defined. The Kyoto Protocol allows the creation of carbon offsets by undertaking renewable power projects in developing countries. However, it has not succeeded in a fundamental shift to renewable power, (see chapter 10). There are no significant signs of such a shift either in the United States or internationally. This chapter will look at both renewable power and conservation investments within the schemes of carbon control.

The Pivotal Role of Carbon Offsets

Offsets are the fungible currency of carbon. A GHG offset can be defined as the reduction, removal, or avoidance of GHG emissions from a specific unregulated eligible project that is used to compensate for GHG emissions that occur elsewhere from a regulated project. Offsets are the alternative compliance mechanism for direct reductions of carbon at regulated carbon emission sources.

The Congressional Research Service of the U.S. Congress called emission offsets "a critical design element."[1] Offsets substantially dampen the price of carbon compliance, by widening the array of compliance options.[2] Traditionally, offsets have offered an incentive for lower-cost compliance options.

The use of offsets for compliance increases supply of credits, which can decrease total costs of compliance by an estimated 71%.[3] Prices paid in global and U.S. markets for the sale of offsets ranged from $1.83 to $306/$CO_2$e, with a volume weighted average price of $6. Of the projects tracked that produced offsets, only 23 of the 211 in the United States occurred in the 10 RGGI states, which is the only place in the United States that their application has regulatory value at the date of this book.

Their impact can be significant. At the macro level, RGGI allows offsets to satisfy 3.3–10% of legal carbon compliance obligations, depending on allowance trade pricing thresholds.[4] This may seem like a minor percentage until one realizes that RGGI requires no state reduction in carbon between 2009 and 2015, and then a cumulative 10% reduction by 2018. Under certain pricing threshold contingencies, all compliance with carbon reductions could occur away from, and unrelated to, the regulated RGGI power generating facilities. During the next decade, a 10% use of offsets, if permitted by the pricing threshold, would meet the entire share of legal reductions required under RGGI for system compliance.

According to one power industry source, offsets are a "main avenue of compliance," because there is little that can be done at an existing fossil-fuel-fired facility to control CO_2 emissions.[5] Where international offsets are eligible as options to create compliance, it is expected that the potentially lower cost of implementing carbon offsets in developing countries, as opposed to in those developed countries that are regulating carbon, would dominate the early years of offset creation and registration. International-only, as opposed to domestic, offsets are part of both the EU-ETS and Kyoto Protocol. RGGI allows international offsets if the trading price of credits exceeds $10/ton. California will likely follow suit with its pending carbon regulation. California, in fact, has executed a number of carbon agreements with other states or foreign governments, including with Mexico; Sao Paulo, Brazil; the United Kingdom; British Columbia; and the European Union.

Conservation and Renewable Power in the United States and RGGI Schemes

The RGGI program offsets

In the U.S. RGGI program, no offset credit is allowed for any project that has an electric generation component, unless the project sponsor transfers legal rights to the credits to the regulatory agency.[6] The RGGI scheme does not contemplate that renewable energy projects may create offsets for program compliance. Despite controversy over this point, it was believed by the states that renewable energy projects do not themselves diminish CO_2 emissions. Renewable projects do not generate CO_2 emissions, but whether they displace other CO_2-emitting power generation sources, indirectly through substitution, remains in dispute. This debate turns on issues of reliability and the location of the renewable resource.

Moreover, the RGGI Model Rule disallows offset allowances for any offset project that receives funding or other incentives from state renewable energy trust funds or any credits or allowances that would be earned from any other mandatory or voluntary GHG programs.[7] These measures are quite restrictive considering that renewable portfolio standard energy credits are now awarded in more than half the states. Therefore, the RGGI scheme stands apart from other carbon schemes and even from the renewable energy incentive programs that the RGGI states may have otherwise adopted and implemented.

In addition, the RGGI Model Rule implies, albeit with some ambiguity, that energy conservation projects can qualify to generate offsets. The avoidance of burning fossil fuels due to end-use efficiency on the consumers' side of the meter appears to be an eligible offset project. Nonetheless, the question of whether the reduction of CO_2 emissions directly relates to combustion remains unanswered in the RGGI Model Rule, and will have to be determined as states administer their RGGI programs.

For example, fossil-fuel-burning efficiency improvements to the combustion device itself—the furnace or boiler—may qualify as an offset project. Going one step further, it is less transparent in the Model Rule whether the installation of building thermal efficiency measures—which saves CO_2 emissions by making the building use and retain heat more efficiently, and thus requires less operation of existing fossil-fuel-burning equipment even if the equipment itself is not made more efficient—could qualify as an offset project. Finally, could one go two degrees of separation, and make electricity-using appliances on the customer side of the meter more efficient, and, where there is fossil-fuel-fired generation in the regional electricity mix, claim a proportionate reduction in the dispatch and operation of such equipment and resultant diminution of CO_2 emissions from such reduced fuel burning? The Model Rule does not address these specifications.

No credits can be awarded for projects that are required by any local, state, or federal law, regulation, or administrative or judicial order. Thus, retrofits, efficiency improvements, or emission reductions required by regulation or embodied in permits or administratively issued legal consent decrees will not create salable offset credits. Therefore, voluntary reductions of CO_2 emissions at an existing large power plant will not create a salable offset credit unless the unit proactively gets out in front of the curve of progressively tightening regulatory mandates to achieve, verify, register, and receive a salable offset credit prior to such reduction being included in its emission permit limits. Renewable energy projects are delegated by regulation to a separate universe from carbon reduction in the RGGI scheme.

The Waxman-Markey carbon bill

Offsets have been some of the most controversial provisions of the Waxman-Markey carbon legislation in the United States. Offsets have always been part of this bill since it worked its way through the House of Representatives in mid-2009. However, the offset definitions and provisions became a key bargaining point for various interest groups. To obtain enough votes to constitute the narrow 219–212 vote for passage on the House floor, major concessions were made in amendments to the bill that weakened the targets in the legislation, allocated free allowances to a wide variety of energy-related, consumer-related, and trade-sensitive entities, and moved away from prior basic provisions of carbon regulation. There was significant controversy surrounding the allocation of free allowances, the periods of years for such allocations, and the ultimate auction of emission rights to the highest bidders, for the first time in UI.S. or world environmental regulation.

Offsets function as additional currency in the allowance system, in that they function as substitutes for allowances for purposes of regulatory compliance. They become interchangeable regulatory assets for those entities regulated. Offsets are typically defined as carbon emission reductions occurring outside of regulated facilities. Therefore, this encourages the private market to seek out and develop carbon reduction projects in the most cost-effective locations using the most cost-effective technologies eligible under the regulatory scheme for offsets. In addition to creating an alternative mechanism for regulated entities to achieve carbon reductions away from the regulated entity, this creates an alternative profit center for those undertaking activities that produce offsets. The experience in the EU-ETS is that offsets have traded at a discounted price to the trading price of allowances.

The American carbon legislation in the House of Representatives backed away from immediate auction of all allowances under the program. It was agreed that 85% of the allowances in the early years of the program be given away to various carbon emitters, rather than auctioned as initially proposed by President Obama and contained in the original legislation. This was necessary for obtaining legislative support necessary to pass the legislation. There was a political struggle among various large industries and interest groups to receive a share of these free allowances. Rural power cooperatives and various industries received a share of the free allowances, which also can be sold to others in need who did not receive enough free allowances.

Rural interests received a significant share of legislative concessions as part of last-minute amendments to the bill that swelled its original 600-page length to more than 1,400 pages containing a number of exceptions, qualifications, and special-interest programs. It is fair to say that members of Congress had no time to read this inflation of pages that added concessions. Agricultural interests were successful in winning concessions transferring the responsibility for offsets from the Environmental Protection Agency, not favored by farm state representatives, to the U.S. Department of Agriculture, which has a large staff in the farm states devoted to promoting agricultural programs and is thought to favor rural offsets.

The list of offset activities also was broadened, causing some to accuse the bill of now directing a multibillion-dollar windfall to farm interests. A provision inserted by farm interests postpones any accounting of carbon emissions from biofuels, including corn-based ethanol. A mandatory import tariff was also imposed on future imports from countries that do not adopt comparable controls on carbon emissions. When this takes effect in 2020 under the bill, it may violate trade treaties then applicable.

Representative Joe Barton, a Texas Republican, referred to the deal cutting to special interests as "unprecedented" in legislation. Representative Henry Waxman, the bill's sponsor, said, "Tackling hard issues that have been ignored for years is never easy." President Obama noted about this bill, "the right balance between providing new incentives to businesses, but not giving away the store, is always an art." The jury remains out on the lasting long-term quality of this art.

The Kyoto-EU Carbon System and Renewable Projects

The Kyoto Protocol does not operate in such separate carbon and renewable universes as in the RGGI system. The fundamental assumption was that control of carbon under the Kyoto Protocol would result in a transition to renewable energy as an eventual world energy base. CO_2 is the dominant GHG in all countries, and power generation is typically its dominant source, along with the transportation sector. Renewable power has the distinct characteristic of not contributing to CO_2 emissions.

This choice of power generation technologies is very important. Resource economists believe that Asia has fossil fuel reserves enough to last for over 100 years.[8] However, more than 90% of these fossil reserves are coal, the most significant contributor to global warming. Several of these nations, most notably China and India, are already highly dependent on coal as their principal energy source. In 2003 alone, China's oil consumption jumped by nearly a third, domestic coal production increased by 100 million tons, and electricity consumption rose by 15%.[9]

However, there is no mandatory environmental or sustainability assessment in Kyoto CDM CER offset projects or any requirement of public input, which was rejected as an infringement on host country sovereignty.[10] The Kyoto CDM mechanism treats carbon as a global commodity, ignoring its source or location, and encouraging businesses to seek out and exploit the cheapest carbon-reduction technologies, regardless of whether or not they lead to replacement of the power-generating base with renewable alternatives. The impact of CDM projects has not been to promote appropriate renewable investments in developing countries, as much as it has served to create additional credits for Annex I countries (see chapter 14).

The early experience from the EU trading scheme paralleling the Kyoto Protocol illustrates, similarly, that rather than cut fossil fuel use in developed countries, the typical response to date has been to create CER offsets in developing countries that do not significantly utilize renewable resources, which when transferred then increase the entire cap emission quantity of available emissions in developed countries.[11] A map of CDM location by geography is shown in figure 16-1.

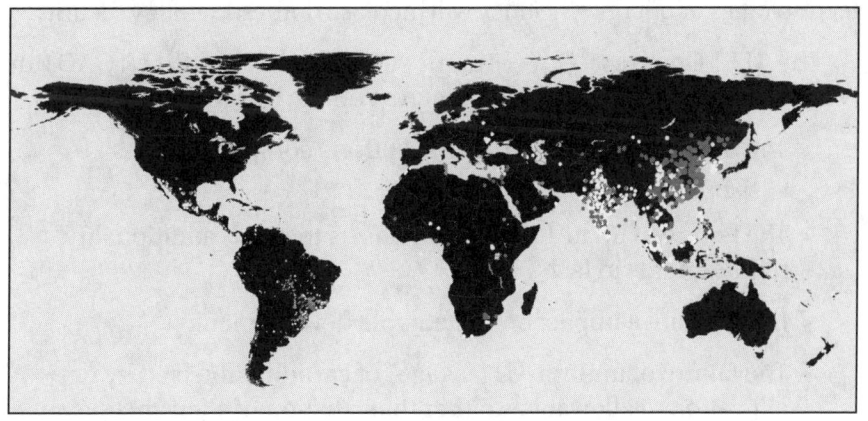

Fig. 16-1. Worldwide location of CDM projects (*Source: UNFCC project activities interactive map*)

Unlimited emission allowance and offset trading is allowed under the EU-ETS and the Kyoto Protocol. Therefore, any party, even if for purposes of speculation, can purchase CDM or RGGI offset credits, even if it does not itself require offsets for its own compliance. The majority of CDM projects, which all must be sited in (non-Annex I) developing countries, have been projects to reduce by-product emissions of HFC-23, a refrigerant (see chapter 5). HFC-23 is not related to the power industry.

An HFC-23 gas mitigation CDM project in a developing country does not shift or promote the power-generating base in either the developing host country or in the country that imports, counts, and applies the CDM CER. HFCs constitute less than 1% of GHGs, but they have received about half of the international investment in mitigation of GHG. Because of the 11,700 times greater CO_2e credits earned from reducing a molecule of HFC emissions, for an investment of €100 million in such CDM projects, it generates CDM credit trading revenues worth about €4.6 billion. When one can reduce HFCs and create CDM offsets for less than $1/CDM credit created, compared to renewable energy projects which can cost $10/ CDM credit created, the market is responding rationally when it favors investments in the former.

World Bank data shows that more than two-thirds of its carbon reduction achievements have involved HFC-23. The issue is that HFC capture and destruction projects do nothing to shift the energy base of the world's economies to sustainable, renewable technologies. The major necessary structural energy sector transformations to low-carbon technologies are not occurring. Deployment of renewable energy generation bases will be required to alter this trend. Renewable energy technologies or energy efficiency will increase domestic energy security.[12]

The U.S. Congress Government Accountability Office (GAO) in November 2008 reported to the Congress on the lesson of the EU-ETS:[13]

- The high cost of producing CDM offsets compared to less costly options

- The failure of up to 40% of CDM offsets to meet "additionality" requirements in fact

- Little positive impact on sustainable development

- The failure of undesired "leakage" of carbon, protected by free allocation of allowances rather than auction of allowances

Can Renewable Power in Kyoto Get *There* from *Here*?

The Congressional GAO concluded that the "evidence indicates that the CDM has had a limited effect on sustainable development."[14] GAO concluded that by encouraging the lowest-cost means for a developer to reduce carbon, the CDM scheme disadvantages measures that contribute to sustainable development. GAO concluded that developing countries that host CDM projects dilute the stringent contribution of CDM programs to sustainable development, because they are competing for these projects to be located in their countries.

Since Kyoto Protocol CDM projects must occur in developing nations, which may or may not have access to sufficient capital, international finance mechanisms, such as those administered by the World Bank and regional development banks, can play a critical role in such projects. World Bank CER acquisition activities have grown from the original Prototype Carbon Fund (PCF) created in 1999, to now 10 additional carbon funds with a capitalization exceeding $2 billion. These can provide upfront financing for CDM carbon reduction projects, and reduce or mitigate regulatory uncertainty and risk of eventual UN Kyoto CDM CER offset credit certification, credit delivery, and political risk. A variety of private insurance products also are available for project-related risks.

It obviously does not fall exclusively on these international development banks. The World Bank is responsible for about one-quarter of the approved protocols certifying CDM offset project development, and influence the lending of regional development banks around the world, national export banks in major developed countries, and private banks, which all underwrite and support infrastructure development. Yet, more than 90% of project financing in developing countries is provided by private banks that have adopted the Equator Principles[15] for lending, and which look for guidance to the World Bank.

Yet, the reality is that one may not get "there" from here. With the RGGI program in the United States excluding renewable energy projects, and the international CDM scheme (used both in the Kyoto Protocol and in the separate EU-ETS carbon regulation systems) including renewable CDM projects as eligible but with relatively few renewable power project realized in either developed or developing counties, there is no fundamental shift in course yet occurring. Even in the Kyoto Protocol, renewable power is allowed, but not required. It is only an option, and one that is proving to be more expensive than exploiting less costly carbon-reduction CDM projects that address other greenhouse gas emissions. Clearly, more affirmative legal incentives in all programs are required to shift the generating base to renewable power or to motivate and credit energy efficiency improvements to create more tradable offsets.

So if the carbon regulatory schemes are not getting there, two basic regulatory options exist. First, the fundamental architecture of the various carbon regulatory schemes in the United States and internationally could be revised to incentivize and/or require more offsets created from renewable power and energy efficiency options. Second, there is the *flip side* of the global warming coin. Instead of trying to limit carbon molecule emissions, one could additionally incentivize or require a certain percentage of renewable power generation in the power supply mix.

While market forces have largely been responsible for encouraging renewable, a new set of strong regulatory requirements for renewable power have been tried in both the United States and Europe, with notable success. However, what has worked in Europe may be unconstitutional when implemented by U.S. states. Nonetheless, approximately 10 states are forging ahead entranced by this policy option, and somewhat oblivious to the important legal issues. There are legal and effective U.S. alternatives to these options.

The next two chapters explore these issues from a policy, regulatory, and legal perspective. Chapters 17 and 18 address the primary European and U.S. affirmative mechanisms to accomplish this shift to renewable power, highlighting feed-in tariffs and renewable portfolio standards. Chapter 19 sets forth the successful model that has been deployed in developing countries to promote renewable energy development. So what follows showcases avenues for successful and legally defensible accomplishment of the flip side of the carbon emission coin: encouraging a shift to renewable power options.

Notes

1   U.S. Congress. Congressional Research Service. 2008. *The Role of Offsets in a Greenhouse Gas Emissions Cap-and-Trade Program: Potential Benefits and Concerns*. CRS Report RL34436, April 4: summary.

2   Ibid., CRS-13; U.S. EPA. 2007. *EPA Analysis of the Climate Stewardship and Innovation Act of 2007*; U.S. EPA. 2008. *EPA Analysis of the Lieberman-Warner Climate Security Act of 2008*. S. 2191, 110th Congress.

3   U.S. Government Accountability Office. 2008. *Carbon Offsets: The U.S. Voluntary Market Is Growing, but Quality Assurances Poses Challenges for Market Participants*. GAO-08-1048. (August): 33.

4   Regional Greenhouse Gas Initiative. 2007. *Model Rule*. Sec. 6.5(a)(3)(i)–(iii).

5   Carbon Control News. 2008. "RGGI Officials Facing Unresolved Questions Over Offset Project Policy." *carboncontrolnews*.com August 18: 7–8.

6   Regional Greenhouse Gas Intitiative, *Model Rule*, sec. 10.3(d)(2).

7   Ibid., sec. 10.3(d).

8   See, International Energy Agency. 2006. *World Energy Outlook 2006*. 88, 115, 127.

9   Yardley, Jim. 2004. China's Economic Engine Needs Power (Lots of It). *New York Times*. March 14: Week in Review, 3.

10 Voigt, Christine. 2008. Is the Clean Development Mechanism Sustainable? *Sustainable Development Law & Policy*. American University, Washington College of Law. (Winter): 15, 20.

11 Ball, Jeffrey. 2007. Kyoto's Caps on Emissions Hit Snag in Marketplace. *Wall Street Journal*. December 3: A-1, 19.

12 Blair, Tony. 2008. *Breaking the Climate Deadlock: A Global Deal for Our Low-Carbon Future*. The Climate Group. 10.

13 U.S. Government Accountability Office. 2008. *International Climate Change Programs: Lessons Learned from the European Union's Emissions Trading Scheme and the Kyoto Protocol's Clean Development Mechanism*. GAO-09-151. (November): 7–8.

14 Ibid., 43.

15 The Equator Principles. www.equator-principles.com/principles.shtml.

17) THE FEED-IN TARIFF FOR RENEWABLE ENERGY: WHERE IT WORKS AND WHERE IT ENCOUNTERS LEGAL IMPEDIMENTS

Ten U.S. states in 2010 are vigorously moving toward implementing a feed-in tariff regulatory mechanism adopted previously by 17 of the European Kyoto Protocol countries as a mechanism to shift to renewable power technologies. However, these state feed-in tariffs could be found to violate the U.S. Constitution, and plunge policy over an immovable legal cliff. For an effective global push against global warming, the ends must not legally be confused with the means, which are not internationally interchangeable under different systems of law. The common goals of reducing the concentration of global warming emissions by substituting renewable energy cannot be implemented with the same uniform tools under different legal systems in the United States and Europe.

Feed-in Tariffs Internationally

Feed-in tariffs are the most widely employed renewable energy policy in Europe and increasingly, the rest of the world.[1] Seventeen European Union countries, plus Brazil, Indonesia, Israel, Korea, Nicaragua, Norway, Sri Lanka, Switzerland, and Turkey, all implemented feed-in tariffs to promote and support renewable energy. In March of 2008, the Kenyan Ministry of Energy proposed the adoption of feed-in tariffs for wind, biomass, and small-hydro resources.

A feed-in tariff establishes a secure contract for wholesale electricity sale of power once a project is in operation, at a set price that results in a rate of return attractive to project investors and developers. The goal is to set the price for power not at its market worth among competitive wholesale power sellers, but at an amount high enough to be attractive to power developers. Feed-in tariff structures are typically either fixed payments based on an electricity generator's cost to produce electricity, or a fixed premium paid above the spot market or wholesale market price of electricity. These fixed payments are embodied in long-term contracts 5–30 years in duration.

These feed-in tariffs increase the price for certain renewable technologies, and the prices ultimately are passed on to retail purchasers of the power, to an amount that is deemed administratively and politically necessary to encourage renewable energy development. These tariffs are set administratively by government regulators, rather than by the power markets. These feed-in tariffs typically may exceed utility-avoided costs, and therefore are justified only by their achieved objective and results, and are not typically accepted ratemaking methodology to minimize prudent generating costs.

Often the fixed-payment feed-in rates and terms are differentiated by technology and are based on the cost of deploying a given renewable energy technology. Feed-in tariffs for sale of renewable power typically decline over time as the high front-end capital costs of renewable energy are amortized and as the number of installed systems increase. Feed-in tariff laws usually also guarantee interconnection for distributed generation and utility-scale projects. Feed-in tariffs have been successful in encouraging significant renewable energy development in nearly all of the countries in which they have been deployed.

As long as a generator feeds in power to the grid, it is guaranteed a long-term contract at the government mandated feed-in price for the renewable energy commodity. A feed-in tariff also can be structured to reflect the benefits that renewable energy sources provide that are not reflected in traditional pricing structures based on fossil fuel resources, including pollution costs, climate change costs, security costs, and future fossil fuel cost uncertainty. Costs of a feed-in tariff are passed on to consumers by purchasing wholesale energy suppliers, and reflect a public policy decision to increase the percentage of renewable electricity sources in use.

For the renewable energy developer, the feed-in tariff decreases investment risk. A feed-in tariff guarantees an investor or developer a long-term contract at a secured price with a return on investment of 8–9%.[2] By contrast, renewable portfolio standard (RPS) policies require developers and investors to secure contracts, which may not be long term, separately for energy and for renewable energy credits (see chapter 18). Finding long-term contracts for two power-related commodities in two different markets injects more risk for investors. One study concluded that capital costs for renewable energy investments are significantly lower in countries using feed-in tariffs than in those countries using policies that create higher risks of future return on investment.[3]

With feed-in tariffs, the government sets the price and guarantees interconnection and contract security, while the market then determines the amount of renewable energy projects put into operation at that level of price. Thus, the government controls the price, and the market reacts to determine the quantity of projects developed in response. The RPS system (see chapter 18) is different: the government specifies by regulation the quantity of renewable power required, and the value of such entitlements is determined by market trades of the regulatory credits earned by that quantity of renewably produced energy. Therefore, there is a fundamental choice of whether a regulatory scheme controls quantity or price, with the market determining the other independent variables associated with renewable energy. Both systems have had success. Germany's feed-in tariff program has created one of the world's largest solar energy markets, and Spain is close behind.

Feed-in Tariff Concepts Developed in the United States

Feed-in tariffs have not been sanctioned historically in the United States. The most prevalent renewable energy policy enacted by states is the renewable portfolio standard (see chapter 18) with a similar renewable energy standard concept in the Waxman-Markey energy legislation. Both are similar to the extent that they only qualify renewable power to the extent that it is actually produced. The feed-in tariff does this by actually linking the renewable subsidy to the price paid for renewable power produced, while the RPS does this by creating a separate tradable renewable attribute that can be traded apart from the power itself.

Some U.S. states have begun to propose legislation and adopt policies similar to European feed-in tariffs. The Solar Electric Power Association issued a report urging utilities to adopt feed-in tariffs, apparently oblivious to the legal pitfalls and ramifications, discussed later in this chapter. As many as 10 states have introduced actual feed-in tariff legislation, while a handful of others are considering feed-in tariff policies.

The federal proposal

In the spring of 2008, and again in 2009, U.S. Congressman Jay Inslee introduced federal feed-in tariff legislation for renewable power that would guarantee uniform interconnection standards, provide for a mandatory 20-year purchase requirement, and set up rate recovery through a national system benefit charge. This proposal would combine feed-in tariffs with RPS and system benefit charge concepts similar to state programs to date (see chapter 18). This proposed legislative scheme allows a 20-year tariff payment at prescribed rates federally established in an amount differentiated by the renewable technology. These payments would be linked to new, federally created renewable energy credits (RECs). Therefore, the federal scheme would co-opt several state concepts. Inslee himself has noted that there are significant barriers to passing a national feed-in tariff statute, including inequalities that could result from how federal funds are allocated to individual states.

State legislative proposals

There are several types of state feed-in tariff incentives proposed and on the horizon, but not yet enacted.

California. The California Public Utilities Commission (CPUC) established the California Solar Initiative. This initiative is a performance-based incentive where solar energy systems can receive a five-year contract worth up to $0.39 per kWh for power sold, approximately 5–10 times the market value of wholesale power. The program is similar to a German-style feed-in tariff, but is shorter in contract term and well below the rates in Germany. The incentive amounts decrease over time after legislative installed solar capacity targets are met. This was for renewable energy systems smaller than 1.5 MW in capacity. Subsequently, the CPUC commissioners accepted the California Energy Commission staff's recommendation that California implement a system of feed-in tariffs for projects up to 20 MW in size.

Hawaii. The legislative session in Hawaii is considering proposals sought to establish 20-year contracts at a rate of $0.40–0.70 per kWh for solar systems up to 20 MW in capacity and only apply to excess electricity sold into the grid from net-metered systems.

Illinois. Illinois considered providing solar generators a net-metering rate of 200% of the retail—not wholesale—rate for electricity for 20-year contracts.

Michigan. Michigan considered a bill to require electric utilities to enter into power purchase agreements for a term of not less than 20 years and to purchase all electricity from eligible electric generators at specified rates, which would be the highest on the North American continent:

- $0.10/kWh for electricity from hydroelectric projects less than 500 kW
- $0.145/kWh for electricity from biogas projects less than 150 kW
- $0.19/kWh for electricity from geothermal projects less than 5 MW
- $0.65/kWh for electricity from rooftop solar installations less than 30 kW
- $0.71/kWh for electricity from solar cladding less than 30 kW
- $0.105/kWh for electricity from commercial wind projects
- $0.25/kWh for electricity from small wind turbines

Minnesota. Minnesota is considering a bill similar to the Michigan proposal, except applying only to Minnesota projects, which raises dormant Commerce Clause implications and Federal Power Act implications (see the following section).

Rhode Island. Rhode Island is considering a bill similar to the Michigan bill offering a 20-year contract at rates that vary depending on the capacity of the generator. All technologies receive rates greater than the avoided cost or market-based power rates.

Florida. The Gainesville, Florida, city commission approved a tariff of $0.32/kWh under Gainesville Regional Utilities' proposed feed-in tariff program, becoming the first feed-in tariff in the United States. This is a municipal utility.

Indiana. Indiana is considering a bill patterned after the Michigan bill.

Vermont. The Vermont Sustainably Priced Energy Enterprise Development Program was amended to allow projects less than 1 MW in capacity to enter into contracts 15 years in length at prices adequate to promote renewable resources. This program *could* be developed into a feed-in tariff if the contract rates are high enough to promote renewable resource development.

Wisconsin. The governor's task force on global warming recommended adopting an advanced renewable tariff for projects 15 MW in capacity and smaller at cost-plus-profit rates. A report developed for the state would set biomass tariffs at $0.15–0.43/kWh, with solar tariffs at $0.43–0.53/kWh, and wind tariffs at $0.09–0.23/kWh. All of these are several-fold higher than the value of generic wholesale power.

Federal Preemption of State Authority for Wholesale Rate Determinations

The Federal Power Act, through sections 205 and 206, empowers FERC to regulate rates for the interstate and wholesale sale and transmission of electricity.[4] In doing so, the act bestows upon FERC broad power to shape the energy market and affect all stakeholders: generators, retailers, and consumers. As discussed also in chapter 14 dealing with carbon allowance auction, by exercising exclusive authority over "just and reasonable" rates and terms, FERC assures wholesale generators of electric power will charge fair rates to retailers and that wholesale generators will receive a fair rate of return, and thus have the incentive to continue to produce and supply power.[5] The act creates a "bright line" between state and federal jurisdiction, with wholesale power sales falling on the affirmative federal side of the line.[6]

As stated in chapter 14, FERC jurisdiction preempts state regulation of wholesale power transactions and prices. The Federal Power Act defines "sale at wholesale" as any sale to any person for resale.[7] FERC also regulates power generation (to a limited degree), power transmission in interstate commerce, and interstate power sales.[8] FERC jurisdiction is plenary and extends to all sales in interstate commerce.[9]

There is no doubt that renewable power sales designed to affect (1) wholesale power transactions and (2) interstate power transactions are subject to exclusive federal jurisdiction; state authority is preempted and stricken. As the United States Court of Appeals for the Ninth Circuit has remarked, and the Supreme Court confirmed, when combined with federal preemption precedent, energy market regulatory reforms have contributed to "a massive shift in regulatory jurisdiction from the states to the FERC."[10]

The Filed Rate Doctrine

If a utility or independent power producer is subject to FERC jurisdiction and regulation, state regulation of the same rate and operational aspects is preempted as a matter of federal law.[11] Principles of preemption require a state regulatory agency to accept and pass through in retail rates all cost items deemed by FERC to be "just and reasonable" or within FERC's jurisdiction even if it has not yet acted on what is otherwise allowed.[12] The so-called filed-rate doctrine holds that state regulatory agencies may not second-guess or overrule on any grounds a wholesale rate determination made pursuant to federal jurisdiction.[13] The Supreme Court in 1986 (and again in 1988, 2003, and 2008) upheld the filed rate doctrine with regard to power pricing.[14]

Pursuant to the filed rate doctrine, the filed federal rate becomes "the legal rate."[15] Outside the regulatory scheme, the filed rate cannot be attacked as the result of improper conduct:[16] "[T]he filed rate doctrine bars all claims—state and federal—that attempt to challenge the terms of a tariff that a federal agency has reviewed and filed."[17]

Feed-in tariff rates above avoided cost or market cost result in at least a temporary, and perhaps longer, increased cost of electricity. And here lies the conundrum: does this conflict with either the requirements of the Public Utility Regulatory Policies Act (PURPA), which are part of the Federal Power Act, or the general rate-setting requirements of FERC under the Federal Power Act? A series of court decisions over the past two decades makes this a very appropriate question under the Supremacy Clause of the United States Constitution.

Are there any exceptions to this? Yes, there is no restriction to two parties agreeing to a particular wholesale price for power sales, as long as they are not required by state regulatory authorities to do so. Similarly, if people choose, rather than being required by regulations, to purchase higher-priced renewable energy to get its renewable characteristics, or in situations involving a net metering option permitted by states, there are exceptions. The following sections examine both.

The green energy limited exemption

Only two very limited exceptions allow utilities legally to pay in excess or to have states mandate that utilities pay in excess of utility avoided costs for renewable energy produced and delivered. The first is if the excess cost is for a green energy program in which utility customers individually voluntarily agree to higher rates covering the costs above the utility's avoided cost.[18] A cost-recovering and appropriately priced green electricity purchase would likely be prohibitively expensive to many consumers, compared to the lesser rates for conventional purchase of electricity. For example, voluntary programs consisting of RPS eligible RECs and future RECs can vary in cost from $0.014 per kWh to $0.50 per kWh in Massachusetts.[19]

The net metering exemption

The second exception applies to a form of metering known as net metering. On March 28, 2001, the Federal Energy Regulatory Commission held that state net metering decisions were not preempted by federal law.[20] In its holding, FERC held that no sale occurs when an individual customer installs distributed generation and accounts for its dealings with the utility through the practice of *netting*, where one ignores individual power sales and looks only at the net sales/purchases of power at the end of each month, and the state chooses to ignore all interim sales of power. It deemed these interim changes of title to power not to constitute a sale.

Oregon has gone even further. Oregon ruled that a customer-generator could hire a third party to own and/or install and operate a self-generation unit on its premises that supplied power behind the meter, and the sale of any such power to the customer or utility was not deemed to be a regulated retail or wholesale sale of power, depending on whether one utilizes 50% wind power through a local utility purchase, or other sources at higher cost.[21] Massachusetts in 2010 adopted a solar photovoltaic REC program, like the one in New Jersey, where failure of a retail power seller to obtain enough such solar power imposes a $0.60/kWh alternative penalty on the deficient seller.[22] The regulated utility is not required in any manner to have to determine who owns net metering facilities. From the sale of power from this net-metered facility, the third-party owner of the renewable generation equipment can still collect RECs (see chapter 18). The ability to quadruple-dip into RECs, net metering, tax incentives, and system benefit trust funds or other subsidies is not uniformly allowed in other states.

In the United States, 80% of the states have electively adopted net metering, which runs the retail utility meter backward when a renewable energy generator puts power back to the grid (see fig. 17–1). Net metering can pay the eligible renewable energy source approximately three or four times more for this power when it rolls backward the retail rate, than paid to any other independent power generators for wholesale power. This is much more than the time-dependent value of this wholesale power to the purchasing utility. The state positions on net metering are set forth in table 17–1.

By turning the meter backward, net metering effectively compensates the power generator at the full *retail* rate for transferring just the *wholesale* energy commodity. While most states compensate the generator for excess generation at the avoided cost or market-determined wholesale rate, as table 7-1 shows, some states have compensated the wholesale energy seller for the excess at the fully loaded, and much higher, retail rate. The excess power, (even if transferred back against the utility's wishes) as long as it nets other power sold to the customer, is not a sale if the state so determines under net metering rules.

Net Metering Rules

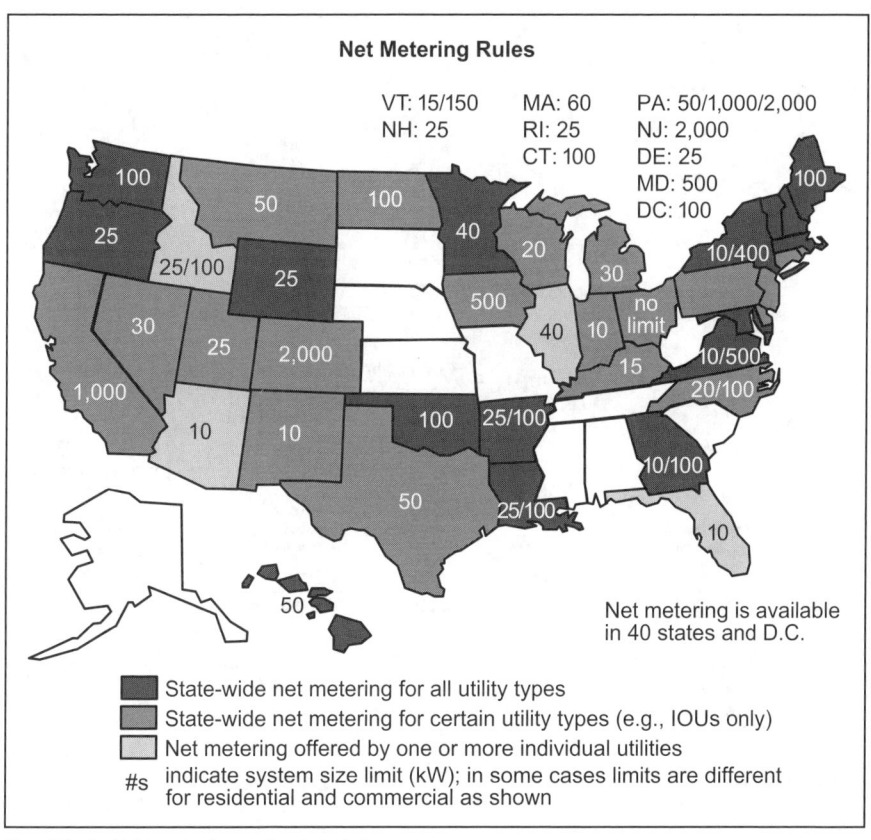

VT: 15/150   MA: 60   PA: 50/1,000/2,000
NH: 25   RI: 25   NJ: 2,000
             CT: 100 DE: 25
                      MD: 500
                      DC: 100

- State-wide net metering for all utility types
- State-wide net metering for certain utility types (e.g., IOUs only)
- Net metering offered by one or more individual utilities
- #s indicate system size limit (kW); in some cases limits are different for residential and commercial as shown

Net metering is available in 40 states and D.C.

Fig. 17–1. Net metering rules by state

It is possible even to game the system with net metering—selling power to the utility at the netted average retail price in off-peak late evening hours when the customer/generator has no need for the power and the utility has surplus power. Other utility ratepayers ultimately will be left to make up the revenue deficit that occurs. Thus, how states treat net energy generation (NEG) is one of the more controversial aspects of net metering.

Table 17-1. State net metering requirements

State	Eligible Technologies	Eligible Customers Limits	Size Limitations	Price	Authorization
Arizona	Renewables & cogeneration		< 100 kW	Excess* purchased at avoided cost	Ariz. Corp. Comm. Decision No. 52345
California	Solar and wind	Residential and small commercial	< 10 kW	Excess purchased at avoided cost; month-to-month carryover allowed w/utility consent	Calif. Pub. Util. Code §2827
Colorado	All resources		< 10 kW	Excess carried over month-to-month	Pub. Svc. Co. of Colo., Advice Letter 1265; Decision C96-901
Connecticut	Renewables & cogeneration		< 50 kW for cogeneration; < 100 kW for renewables	Excess purchased at avoided cost	Dept. of Pub. Util/ Control, Order No. 159
Idaho	Renewables & cogeneration	Residential and small commercial	< 100 kW	Excess purchased at avoided cost	ID PUC Orders Nos. 16025 (1980); 26750 (1997)
Indiana	Renewables & cogeneration		< 1,000 kWh/ month	Excess is "granted" to the utility; No purchase of excess	170 IN Admin. Code §4-4, 1-7
Iowa	Renewables		No size limit	Excess purchased at avoided cost	Iowa Util. Bd., Utilities Division Rule §15.11(5)
Maine	Renewables & cogeneration		< 100 kW	Excess purchased at avoided cost	Me. PU Code Ch. 36, §§1(A)(18), (19), §4(C)(4)
Maryland	Solar	Residential	< 80 kW	Excess carried over to following month	Maryland Art. 78, §54M
Massachusetts	Renewables & cogeneration		< 60 kW = Class I Between 60 kW and 1 MW = Class II Between 1-2 MW = Class III	Excess purchased at avoided cost	Mass. Gen. Laws c. 164, §1G(g); D.T.E. Order 97-111 Note: < 30 kW 220 CMR §8.04(2)
Minnesota	Renewables & cogeneration		< 40 kW	Excess purchased at "average retail utility energy rate"	Minn. Stat. §261B.164(3)
Nevada	Solar and wind		< 10 kW	Excess purchased at avoided cost; annualization allowed	Nev. R. Stat. Ch. 704
New Hampshire	Solar, wind & hyrdo		< 25 kW	PUC may require 'netting' over 12-month period; retailing wheeling allowed for up to 3 customers	
New Mexico	Renewables, fuel cells, micro turbines		< 1,000 kW	Excess credited to following month; unused credit is granted to utility at end of 12-month period	NM PUC Order 2847 (11/30/98)

CHAPTER 17 • THE FEED-IN TARIFF FOR RENEWABLE ENERGY

State	Eligible Technologies	Eligible Customers Limits	Size Limitations	Price	Authorization
New York	Solar	Residential	< 10 kW	Excess credited to following month; unused credit is granted to utility at end of 12-month period	NY Public Service Stat. §66-j
North Dakota	Renewables & cogeneration		< 100 kW	Excess purchased at avoided cost	N.D. Admin. Code §69-09-07-09
Oklahoma	Renewables & cogeneration		< 100 kW and annual output < 25,000 kWh	Excess is "granted" to the utility; no purchase of excess	Ok. Corporations Comm. Schedule QF-2
Pennsylvania	Renewables		< 50 kW	Excess purchased at wholesale rate	PECO Rate R-S, Supp. 5 to PA Tariff PUC No. 2, Page 43A
Rhode Island	Renewables & cogeneration		< 25 kW for larger utilities; < 15 kW for smaller utilities	Excess purchased at avoided cost	PUC Supp. Decision and Order, Docket No. 1549
Texas	Renewables		< 50 kW	Excess purchased at avoided cost	Texas PUC, Rule §23.66(f)(4)
Vermont	Solar, wind, fuel cells using renewable fuel, anaerobic digestion	Residential, commercial, and agricultural customers	<15 kW, except < 100 kW for anaerobic digesters	Excess carried over month-to-month; any residual excess at end of year is "granted" to the utility	Reuse of Net Metering, VT. PSB Docket No. 6181 (April 21, 1999)
Washington	Solar, wind, and hydropower		< 25 kW	Excess credited to following month; unused credit is granted to utility at end of 12-month period	
Wisconsin	All resources	All retail customers	< 20 kW	Excess purchased at retail rate for renewables, avoided cost for non-renewables	Pub. Svc. Comm. Schedule PG-4
Connecticut	Solar, wind, hydro, fuel cell, sustainable biomass	Residential	No size limit	Not specified	CT Public Act 98-28 (1998)
Illinois (pending)	Solar and wind	All retail customers	< 40 kW	Excess carried over month-to-month; any residual excess at end of year is purchased at avoided cost	Ill. Legis. S.B. 1228
Maine	Renewables or other applicable technology		< 100 kW	Excess carried over month-to-month; any residual excess at end of 12-month period is eliminated	Me. PU Code Ch. §313 (1998); PUC Order No. 98-621 (December 19, 1998). [35-A MRSA §3210(2)(C)]
Puerto Rico (pending)	Renewables	Residential	< 50 kW	Excess carried over month-to-month; any residual excess at end of year is purchased at avoided cost	

* "Excess" refers to the "net excess generation" of electricity by the customer-generator (i.e., generation exceeds consumption) during the billing period.

Key State Efforts Constitutionally Stricken: The California Cases

The Ninth Circuit: *Independent Energy Producers*

In *Independent Energy Producers Association v. California Public Utilities Commission*, the California state utility commission authorized utilities to monitor qualifying facilities (QFs) established pursuant to the PURPA amendments to the Federal Power Act, to determine whether they met federal efficiency standards.[23] In addition to allowing monitoring, the state commission allowed the utility to suspend payment to the wholesale power selling QF if the utility found that the QF did not comply with federal standards. The utility was authorized by the state of California to substitute a lower, alternative rate of only 80% of the avoided cost rate in the event that it determined that the QF did not comply.

On review, the federal court of appeals held that a program where the state determined the ultimate renewable energy QF status was preempted by federal law. The court also stated that the rate paid by utilities for electricity must be determined by calculating the avoided cost that the utility itself would pay if it had to purchase wholesales electricity outside the QF contract price. Attempts by states indirectly or directly to promote higher wholesale energy prices for certain renewable energy projects have been stricken by the courts. Promotion of certain types of renewable fuels for power supply, via a price preference above and beyond the FERC-established market price of other wholesale power transactions, was held inconsistent with the Federal Power Act and stricken. States could not tinker with the wholesale price of power in their states.

The FERC backstop: *Southern California Edison Company, San Diego Gas & Electric*

FERC also refused to sanction a higher California price for renewable power supply.[24] Under the filed rate doctrine, any dispute about these matters may not be arbitrated by the state, but is reserved exclusively to federal authority.[25] The California Public Utilities Commission ordered two of its investor-owned and regulated utilities, Southern California Edison and San Diego Gas & Electric, to sign long-term fixed-price contracts with PURPA qualifying facility renewable power projects to purchase electricity at prices that were competitive with other renewable energy prices, but nonetheless in excess of the utilities' avoided cost, which reflected all market wholesale prices. Edison had wholesale electricity supply options available for $0.04 per kWh or less (lower than renewable energy costs), while the California PUC required renewable QF contracts bid as high as $0.066 per kWh to be executed by the utility.

The case went to FERC on challenge. FERC ruled that under PURPA, states have broad powers under state law to direct the planning and resource decisions of regulated retail utilities under their jurisdictions. However, the FERC made it clear that PURPA does not permit either the FERC or the states in their implementation of PURPA, as part of the Federal Power Act, to require a purchase rate that exceeds the utilities' avoided cost or market-based rate for all wholesale power collectively. The avoided cost process must reflect prices available from all wholesale sources able to sell to the utility, whose avoided cost is thereby determined. This concern does not ameliorate over time: FERC further stated that as the electric utility industry becomes increasingly competitive, the need to ensure that the states are using procedures which ensure that QF rates do not exceed avoided cost becomes even more critical. This language foreshadowed the FERC Energy Policy Act of 2005 regulations in 2006 that substitute general market-based grid prices in functional wholesale markets for QF purchasing requirements in Day-2 markets.[26]

Notwithstanding this experience, California in 2009 moved forward toward implementing a feed-in tariff. The initial steps drew some immediate legal challenges from some of the regulated electric utilities. In defense, the California attorney general, Jerry Brown, defended feed-in tariffs as defensible either under PURPA or as part of renewable energy credits in California. On this first defense, using the state authority to set PURPA avoided cost rates, the attorney general claimed that, with the RPS program, an increasing share of incremental power will come from renewable energy, and therefore the cost of renewable energy is the avoided cost. The attorney general also claimed that the need to deal with global warming created a new reality that gave California a longer legal leash. The problem with this logic is that it is inconsistent with either this FERC or United States Court of Appeals for the Ninth Circuit precedent holding that a state in a regulatory mode cannot set the wholesale price nor set PURPA avoided cost with reference only to a portion of the wholesale market.

In claiming that the California RPS program provided California authority to set wholesale power prices, the attorney general leaned on the legally correct perspective that states had control over RECs. This is true, but RECs are not related to the wholesale power market. RPS requirements are imposed on retail utilities as a condition of their license to operate. However, with the regulatory step to set wholesale power sale requirements using feed-in tariffs, such an action is wholly unrelated to the retail market that states are allowed to regulate. In claiming the authority to set differentiated above-market wholesale power feed-in tariffs, a state is crossing the line between allowed state retail authority and disallowed wholesale jurisdiction to require certain wholesale rates.

Ninth Circuit redux: *Public Utility District No. 1 of Snohomish County Washington*

A final key court decision on whether preferential wholesale power rates can be enacted for certain preferred renewable power technologies was rendered by the United States Court of Appeals for the Ninth Circuit[27] and affirmed by the U.S. Supreme Court in 2008.[28] While this decision was affirmed in part by the U.S. Supreme Court on appeal, and remanded to FERC for more clarification or explanation, it was not overturned. State law on power terms and prices are not allowed to preempt federal wholesale term and rate determinations by layering on additional requirements not contained in federal law.[29] FERC not only has exclusive authority unaffected by any state actions over wholesale power markets, but FERC has an ongoing obligation to continually monitor and police these markets against state interference.[30]

As the court of appeals has remarked, and the Supreme Court confirmed, when combined with federal preemption precedent, energy market regulatory reforms have contributed to "a massive shift in regulatory jurisdiction from the states to the FERC." The wholesale price determination is reserved exclusively to federal authority.[31] The Federal Power Act precludes all state regulation of interstate wholesale power transactions, which could include preclusion of state determination of wholesale feed-in tariffs for certain sources of wholesale power.[32]

There are two issues here. First, this could call into question states establishing feed-in tariffs at all, since these are wholesale prices for power sale and are absolutely preempted from any state jurisdiction. They are exclusively within federal jurisdiction. States cannot cross this "bright line." Second, there is ample legal precedent in the United States, discussed earlier, preventing rates set above the wholesale market price for particular favored types of power supply of any kind. This is a U.S. wholesale power market operating at fair and reasonable rates to all stakeholders, rather than a subsidy system for certain technologies that states may favor at a particular point in time.

So what works in Europe and elsewhere—feed-in tariffs—may run afoul of the U.S. constitutional separation of jurisdiction, when implemented by the states. Instead, more than half the U.S. states have implemented renewable portfolio standards. These are the alternative to feed-in tariffs, and generally have been regarded as a success. They are quite varied and can be exotic, and are examined in the next chapter.

Notes

1 See Rickerson, Wilson and Bob Grace. 2007. *The Debate over Fixed Price Incentives for Renewable Electricity in Europe and the United States: Fallout and Future Directions*. Heinrich Böll Foundation. February.

2 Seager, Ashley. 2007. Germany Sets Shining Example in Providing a Harvest for the World. *The Guardian*. July 23.

3 Held, A., M. Ragwitz, C. Huber, G. Resch, T. Faber, and K. Vertin. 2007. *Feed-in Systems in Germany, Spain, and Slovenia: A Comparison*. Karlsruhe, Germany: Fraunhofer Institut für Systemtechnik und Innovationsforschung.

4 *United States Code*, title 16, sec. 824d and 824e.

5 *Snohomish County Public Utility District No. 1 v. FERC*, 471 F.3d 1053, 1058 (9th Cir. 2006)

6 *Snohomish County Public Utility District No. 1 v. FERC*, 471 F.3d 1053, 1066 (2006), affirmed in part and reversed in part on appeal under the name *Morgan Stanley Capital Group v. Public Utility District No. 1 of Snohomish County*, 128 S. Ct. 2733 (2008). For a discussion of the California and Western energy crisis that spawned this litigation, see Ferrey, Steven. 2004. Soft Paths, Hard Choices: Environmental Lessons in the Aftermath of California's Electric Deregulation Debacle. *Virginia Environmental Law Journal*. 23 (2): 251.

7 *United States Code*, title 16, sec. 824(d).

8 *United States Code*, title 16 sec. 824(a). Federal regulation extends only to those matters which are not subject to regulation by the states.

9 *Northern States Power Co. v. Minnesota Public Utilities Commission*, 344 N.W.2d 374 (Minn. 1984), *cert. denied*, 467 U.S. 1256 (1984).

10 *Snohomish County Public Utility District No. 1 v. FERCn*, 471 F.3d at 1067; *Entergy Louisiana, Inc., v. Louisiana Public Service Commission*, 539 U.S. 39 (2003).

11 See, *Arkansas Power & Light Co. v. Federal Power Commission*, 368 F.2d (8th Cir. 1966); *Nantahala Power & Light Co. v. Thornburg*, 476 U.S. 953 (1986); *Appeal of New England Power Co.*, 424 A.2d 807 (1980).

12 *Appeal of Sinclair Machine Products, Inc.*, 498 A.2d 696 (N.H. 1985).

13 However, the Supreme Court has determined that Congress, in enacting the Federal Power Act, intended to vest exclusive jurisdiction in the FERC to regulate interstate wholesale utility rates. *Federal Power Commission v. Southern California Edison Co.*, 376 U.S. 205, 216 (1964); *Narragansett Electric Co., v. Burke*, 119 R.I.

559, 381 A.2d 1358, 1361 (1977), *cert. denied*, 435 U.S. 972 (1978). In this case, federal preemption of state discretion on retail rate pass-through of wholesale rate established pursuant to federal jurisdiction.

14 See, *Nantahala Power & Light Co. v. Thornburg*, 476 U.S. 953, 963 (1986). "This Court has held that the filed rate doctrine applies not only to the federal-court review at issue in Montana-Dakota, but also to decisions of state courts." *Mississippi Power & Light Co. v. Mississippi ex rel. Moore*, 487 U.S. 354 (1988). Filed rate doctrine applies without exception to state regulation of interstate holding companies; *Entergy Louisiana, Inc., v. Louisiana Public Service Commission*, 539 U.S. 39 (2003). No residual prudency power of the states to alter federal rate or term even where matters left to the discretion of the regulated entity.

15 *Keogh v. Chicago & Northwestern Ry. Co.*, 260 U.S. 156, 163 (1922); *Montana-Dakota Utilities Co. v. Northwestern Public Service Co.*, 341 U.S. 246, 251 (1951).

16 *Arkansas Louisiana Gas Co. v. Hall*, 453 U.S. 571, 578 (1981).

17 *People of California ex rel. Lockyer v. Dynegy*, 375 F.3d 831, 853 (9th Cir. 2004); *Transmission Agency of Northern California v. Sierra Pacific Power Co.*, 295 F.3d 918, 929 (9th Cir. 2002).

18 The Florida Public Service Commission found that a green pricing program does not violate PURPA and its implementing rules. However, the Florida PSC made it clear that the commission did not answer the question of whether circumstances might exist where prices in excess of avoided cost could be borne by the general body of ratepayers, or the question of the amount the utility or its green electricity customers could pay. See, *re Florida Power and Light Company*, 219 P.U.R. 4th 46 (2002).

19 NSTAR. NSTAR Green. http://www.nstar.com/residential/customer_information/nstar_green/nstar_green.asp.

20 See, MidAmerican Energy Company, 94 FERC, par. 61, 340; 2001 FERC LEXIS 630. In March 2001, MidAmerican Energy Company challenged before FERC the state of Iowa's regulations directing MidAmerican to interconnect with three "Alternate energy facilities and to offer net billing arrangements to those facilities."

21 Oregon Public Utility Commission. 2008. Honeywell International and Pacific Power, Oregon Public Utility Commission Order No. 08-388, July 31 (interpreting ORS 757.300).

22 Massachusetts Department of Energy Resources. *Solar RPS Carve-Out*. http://www.mass.gov/Eoeea/docs/doer/renewables/solar/MA%20Solar%20RPS%20Carve-Out%20-%20Price%20Support%20Mechanism%20-%20Stakeholder%20Mtg%20102309%20DOER.pdf.

23 *Independent Energy Producers Association v. CAlifornia Public Utilities Commission*, 36 F.3d 848 (9th Cir. 1994).

CHAPTER 17 • THE FEED-IN TARIFF FOR RENEWABLE ENERGY   245

24 Southern California Edison Co. and San Diego Gas & Electric Co., FERC Docket Nos. EL95-16-000 and EL95-19-000, 70 FERC par. 61,125 (1995), requests for reconsideration denied, 71 FERC par. 61,269 (1995). Holding that the costs of renewable energy not to exceed the market or bid price of all other sources of energy makes ratepayers indifferent as to the procurement of wholesale power.

25 *Mississippi Power & Light Co. v. Mississippi ex rel. Moore.*

26 See, Code of Federal Regulations, title 18, part 292, Docket No. RM06-10-001; PURPA. 2007. Ne*w PURPA Section 210(m) Regulations Applicable to Small Power Production and Cogeneration Facilities.* Order No. 688-A, issued June 22.

27 *Public Utility District No. 1 of Snohomish County Washington v. FERC*, 471 F.3d 1053, 1066 (9th Cir. 2006).

28 *Morgan Stanley Capital Group v. Snohomish County Public Utilities District No. 1*, 128 S. Ct. 2733 (2008). The U.S. Supreme Court in its decision criticized the reasoning of the United States Court of Appeals for the Ninth Circuit decision, but nonetheless agreed with and upheld that FERC has exclusive authority, and responsibility, to review long-term power crises, wholesale market manipulation by a party to the power sale contract that would negate existing contract protections, and wholesale rates.

29 *Granite Rock Co. v. California Coastal Commission*, 768 F.2d 1077 (9th Cir. 1985).

30 *Snohomish County Public Utility District No. 1 v. FERC*, 471 F.3d at 1067.

31 *FERC v. Mississippi*, 456 U.S. 765.

32 *Nantahala Power & Light Co. v. Thornburg*, 476 U.S. 953 (1986); *Mississippi Power & Light Co. v. Mississippi ex rel. Moore*, 487 U.S. 354, 371 (1988); *accord Mississippi Industries v. FERC*, 808 F.2d 1525, 1535-49 (D.C. Cir. 1985), *cert. denied*, 484 U.S. 985 (1985).

18 RENEWABLE PORTFOLIO STANDARDS FOR RENEWABLE POWER

RPS Design and Contours

A renewable portfolio standard (RPS) requires certain electricity sellers and/or buyers to maintain a predetermined percentage of renewable energy credits (RECs) produced from the operation of designated clean resources in their wholesale supply mix. RPS programs transfer the risks and benefits of achieving a percentage of renewables to the private sector. REC programs exist in Belgium, Italy, the Netherlands, Sweden, the United Kingdom, and more than half the states in the United States.

RECs are intangible government-created financial assets that can be traded. The RECs have a separate existence from the power production they are associated with, and can be traded or banked depending on the requirements of the issuing and accepting regulatory programs. RECs function both as climate change mitigation and adaptation mechanisms, since they shift the power generating base to renewable power technologies. Trading can involve direct purchase of certificates, or trading derivatives associated with underlying RECs in secondary markets. The trading of RECs is neither a good nor a service, but a financial attribute that is traded through service.

The key to making the portfolio requirements work is to establish trading schemes for portfolio obligations. The standards become self-enforcing as a condition of retail sale licensure. The advantages of a portfolio standard are that it does not subsidize any particular technology or locus of that technology, and does not disturb in any way the federally established wholesale power price. Resource portfolio requirements can be applied under any wholesale or retail competition, or under traditional utility regulation, without placing any entities at a disadvantage.

State program designs vary for RPS programs as to the following:

- Energy versus capacity obligations
- Single-tier or multi-tier credit determinations
- The duration of purchase obligations
- Whether there are requirements for resource diversity
- Whether there are incentives for resource or technology diversity
- Whether all default service providers must participate
- Geographic eligibility for credits
- Whether there is differentiation by type of renewable resource
- Whether credits can be earned only by new renewable generation units or also by pre-existing units
- The definition of new or incremental generation, where that is applicable
- How multi-fuel facilities are categorized
- How off-grid resources are categorized
- Whether distributed generation on the customer side of meters is eligible

Half of the U.S. states have enacted RPS programs to promote renewable energy power production. Half of that half of the U.S. states employ differentiated tiers of RECs, serving various functions for those tiers:

- Some states distinguish tiers by the vintage for the creation of the REC[1]
- Some states designate tiers by type of technology of renewable resource so as to be able to promote a certain technology[2]
- Some states create technology set-asides or bands of technology[3]
- Other states have only a single type of REC regardless of technology, evidenced by a single tier, with only new construction renewable energy projects eligible;[4] other states have a single tier that allows both new and existing projects to qualify[5]

This creates myriad variations on state RPS models.

RPS State Variations and Results

Resources promoted

Twenty-five states plus the District of Columbia had RPS programs in 2009; four additional states have nonbinding RPS goals. These mandatory RPS programs cover 46% of nationwide retail electricity sales.[6] RPS programs were initially created in states that had restructured and/or deregulated their retail power markets after 1997; however, over time, half of the RPS programs were in traditionally regulated and retail supply monopolized states. The early RPS states and those adopting renewable energy trust funds, another subsidy mechanism employed by some states, are illustrated in table 18–1.

The history of RPS programs is relatively recent. In 1991 Iowa established the first renewable portfolio standard. By the end of 2009, more than half the states and the District of Columbia had enacted RPS policies ranging from requirements for as much as 2% to 40% of power from renewable energy achieved over time. Among the most populous states, California has a 33% RPS target by 2020 and New York has a 25% target by 2013.

The RPS programs in the states are very different in terms of what qualifies. Most states allow solar, wind, biomass, and landfill gas resources to qualify in RPS programs; states are less consistent regarding eligibility for biogas, municipal solid waste (MSW), geothermal, all hydro resources, fuel cells, and ocean tidal renewable resources to qualify. Some states count cogeneration, while Pennsylvania includes coal gasification and non-renewable distributed generation. Resource eligibility in state RPS programs has expanded beyond traditional renewables, with three states now allowing demand-side energy efficiency to meet at least a portion of their RPS requirement. Some states set standards based on a percentage of installed capacity, while other states set standards based on a percentage of total electricity sales.

Table 18-1. Portfolio standards and trust funds in early adopter states

State Name	Renewable Energy Trust Fund	Portfolio Standards
Arizona	X	X
California	X	
Colorado		X
Connecticut	X	X
Delaware	X	
Hawaii		X
Illinois	X	
Iowa		X
Maine		X
Maryland		X
Massachusetts	X	X
Minnesota	X	X
Montana	X	
Nevada		X
New Jersey	X	X
New Mexico		X
New York	X	
Ohio	X	
Oregon	X	
Pennsylvania	X	X
Rhode Island	X	
Texas		X
Vermont		X
Wisconsin	X	X

RPS programs have had an impact. Over 50% of the non-hydro renewable capacity additions in the United States from 1998 through 2007 occurred in states with RPS programs; 93% of these additions came from wind power, 4% from biomass, 2% from solar, and 1% from geothermal resources.[7] In those states that have RPS programs, more than 90% of renewable energy additions (and more than 80% of average capacity supplied) are from wind power, with biomass a distant second and limited geothermal resource development. It is estimated that 60–90% of RPS-driven renewable energy capacity additions going forward will be wind power projects. It has been estimated that RPS motivated approximately 45% of the 4,300 MW of wind power installed in the United States between 2001 and the end of 2004, while system benefit charges and trust funds supported an additional 15% of this capacity addition.

In about half of the RPS programs, solar energy installations are being encouraged in a variety of ways. Several states also reward rebates to customers who install solar systems. Solar-specific RPS designs in about a dozen states and Washington, D.C., include solar or distributed generation set-asides for a percentage of eligible projects. These set-aside policies have already supported 102 MW of solar photovoltaics and 65 MW of solar-thermal electric capacity by 2007. Roughly 6,700 MW of solar capacity would be needed by 2025 to fully meet existing set-aside requirements.

The State Checkerboard

Various nuances exist in RPS programs with idiosyncrasies state by state (fig. 18-1). In some cases where RECs have shorter life spans, they can be banked from one year to the next to meet a certain percentage of the next year's annual requirement. RECs for RPS compliance have different longevities and shelf lives. The shelf life of a renewable energy credit is as short as three months in New England and as long as four years in Nevada and Wisconsin. Massachusetts allows banking for up to two years to meet up to 30% of annual requirements, while Delaware, Maryland, and the District of Columbia allow a three-year banking period; California allows indefinite banking.

Eligible project renewable technologies state by state are set forth in table 18-2. Some states allow credits to be traded, while other states do not.

States employing RPS treat generation facilities on the customers' side of meters differently. While Massachusetts and Rhode Island only allowed these resources to earn RECs if they are located within the respective state, Connecticut allows such facilities to earn credits when situated elsewhere in the New England region. Electronic renewable energy credit tracking systems are in place in New England, the PJM interconnection, Texas and Wisconsin, and the Midwestern and Western grid regions. Large portions of the South, outside of Texas, do not have the ability to track RECs. Because the definitions of RECs created under various state programs differ, there is significant geographic limitation in cross-market REC trading and liquidity.

252 UNLOCKING THE GLOBAL WARMING TOOLBOX

Renewables Portfolio Standards

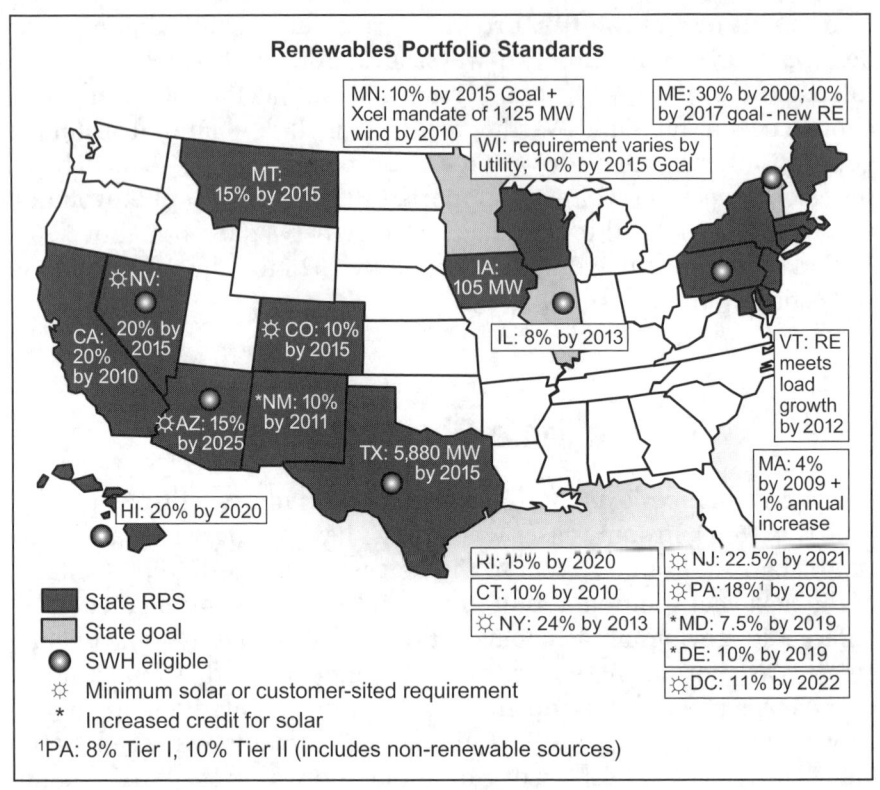

- MN: 10% by 2015 Goal + Xcel mandate of 1,125 MW wind by 2010
- ME: 30% by 2000; 10% by 2017 goal - new RE
- WI: requirement varies by utility; 10% by 2015 Goal
- MT: 15% by 2015
- IA: 105 MW
- ☼ NV:
- CA: 20% by 2010; 20% by 2015
- ☼ CO: 10% by 2015
- IL: 8% by 2013
- VT: RE meets load growth by 2012
- ☼ AZ: 15% by 2025
- *NM: 10% by 2011
- TX: 5,880 MW by 2015
- MA: 4% by 2009 + 1% annual increase
- HI: 20% by 2020
- RI: 15% by 2020
- ☼ NJ: 22.5% by 2021
- CT: 10% by 2010
- ☼ PA: 18%¹ by 2020
- ☼ NY: 24% by 2013
- *MD: 7.5% by 2019
- *DE: 10% by 2019
- ☼ DC: 11% by 2022

Legend:
- ■ State RPS
- ▨ State goal
- ● SWH eligible
- ☼ Minimum solar or customer-sited requirement
- * Increased credit for solar

¹PA: 8% Tier I, 10% Tier II (includes non-renewable sources)

Fig. 18–1. Renewable portfolio standards by state

Noncompliance penalties vary in each state. Average RPS compliance in 2006 was 94%.[8] Alternative compliance payments of more than $18 million were paid in 2006; financial penalties have been applied in two states. The noncompliance or alternative payment penalty is around $0.05/kWh in California, Connecticut, Washington, Rhode Island, Maine, and Massachusetts. The penalty is lower in other states (although New Jersey and New Hampshire have equally high penalties for noncompliance with Class I RECs emissions).

Table 18–2. Renewable resources as defined in state statutes

State	Solar	Wind	Fuel Cell	Methane/Landfill	Biomass	Trash-to-Energy
Arizona	X	X			X	
California	X	X		X	X	X
Connecticut	X	X	X	X	X	X
Iowa	X	X	X		X	
Illinois	X	X			X	X
Maine	X	X	X		X	X
Maryland	X	X	X	X	X	
Massachusetts	X	X	X	X	X	X
Minnesota		X			X	
Nevada	X	X	X			
New Jersey	X	X	X	X	X	X
New Mexico	X	X	X	X	X	X
New York	X	X				X
Oregon	X	X		X		X
Pennsylvania	X	X		X	X	X
Rhode Island	X	X		X	X	X
Texas	X	X		X	X	X
Wisconsin	X	X	X		X	X

State	Hydro	Tidal	Geothermal	Photovoltaic	Dedicated Crops
Arizona		X		X	
California	X		X	X	
Connecticut	X			X	
Iowa				X	
Illinois	X			X	X
Maine	X	X	X	X	
Maryland		X	X	X	
Massachusetts	X	X		X	X
Minnesota					
Nevada			X	X	
New Jersey	X	X	X	X	
New Mexico	X	X	X	X	
New York	X	X	X	X	
Oregon	X	X	X	X	X
Pennsylvania	X		X	X	X
Rhode Island	X			X	
Texas	X	X	X	X	
Wisconsin	X	X	X		X

Note: Photovoltaic is included within solar in some states; methane and or trash-to-energy may be included within a broad definition of "biomass."

The required percentage of energy delivered from renewables can be deceiving depending upon whether preexisting renewable resources are eligible to be counted. Maine is at the 30% level, but because it allows both hydroelectric facilities and preexisting facilities to count, compliance has always been achieved by retailers. A 2007 amendment to the Maine RPS program now requires 10% of renewable power to be 10% of capacity by 2017, starting at 1% in 2008 and increasing by 1% annually. The penalty for noncompliance is set at $57.12/MWh in 2007, which will rise with inflation, and can be waived by the PUC. Failure to comply can result in license revocation or other financial penalties. Rhode Island requires a 3% portfolio standard for renewable energy starting in 2007, rising to 16% of the portfolio in 2020. An alternative compliance payment can be made of $50/MWh in 2003 dollars, escalating, in lieu of meeting the portfolio standard. Buyers also can "bank" renewable certificates for up to two years for future use. Connecticut will recognize RPS credits from other New England states in the New England Power Pool (NEPOOL) system until 2010, and thereafter will additionally recognize credits from New York, Pennsylvania, New Jersey, Maryland, or Delaware if it is determined at that time that their RPS program standards are similar to those of Connecticut. The details of northeastern state RPS programs are illustrated in table 18–3.

The Value of Renewable Energy Credits and Offsets

The prices of traded RECs have been relatively high in three states: (in highest order) Massachusetts, Connecticut (for Class I RECs), and Rhode Island. Other states have had significantly lower RECs trading prices to date. In most states, supply exceeds the demand for RECs, and the prices have trended as low as about 10% of those in the three highest states.

The price impact of RPS-mandated renewable energy project generation has been estimated to range between a 0.1% increase in retail rates (in Maine, Maryland, New Jersey, and New York) up to a 1.1% retail rate impact in Massachusetts.[9] In 2005, Massachusetts collected $19.6 million in alternative compliance payments under its RPS system, and nearly $17.8 million in 2006.[10] Now, there is a national move to such a renewable credit system, as embodied in the Waxman-Markey federal legislation. A utility RPS charge of only $0.001/kWh would raise $2 billion annually if imposed across all retailed power in the United States.

Table 18–3. Seven northeastern state RPS requirements

State	Requirement	Technology Eligibility
Connecticut 3 Classes	Class I technologies: 1% in 2004 +0.5%/yr; to 2% by 2006 +1.5%/yr; to 5% by 2008; +1%/yr to 7% in 2010 and thereafter Class I or II technologies: 3% in 2004 and thereafter	Class I: solar, wind, landfill gas, new (post 7/1/03) run of river hydro (<= 5 MW), fuel cells, ocean thermal, wave or tidal, low-e RE conversion tech., low NO_x emitting, sustainable biomass. (Biomass facilities with quarterly avg. NOx emission rate <= 0.075 lbs. per MMBTU. Existing (pre 7/1/03) biomass facilities <= 500 kW are exempt from NO_x emission requirement.) Class II: MSW, existing (prior to 7/1/03) run of river hydro (<= 5 MW), other biomass (facilities must have quarterly avg. NO_x emission rate <= 0.2 lbs. per MMBTU)
Maine	30% of sales in 2000 (start of competition) and thereafter as a condition of licensing	Fuel cells, tidal power, solar, wind, geothermal, hydro, biomass, and MSW (under 100 MW). High efficiency cogen. systems of unlimited size.
Maryland	Tier 1 Renewables: 1% in 2006, increasing 1% biannually to 7% in 2018, increasing to 7.5% in 2019, and thereafter Tier 1 or 2 Renewables: 2.5% 2006-2018	Tier I: solar, wind, biomass, landfill gas, geothermal, ocean, fuel cells (renewable sources only), and small hydro (< 30 MW) Tier 2: hydro, MSW, and incineration of poultry litter
Massachusetts	1% of sales from new renewables by 2003 +0.5%/yr. to 4% in 2009; +1 % per year thereafter until date determined by Division of Energy Resources (moved to 3 classes of RECs in 2009, including for new, vintage, and solar technologies)	Solar, wind, ocean thermal, wave, tidal, landfill gas, and low-emission advanced biomass beginning commercial operation or representing increase in capacity at existing facility after 12/31/97. Hydro and MSW qualify as existing and are not eligible.
New Jersey 3 Classes	Class I or II Technologies: 2.5% by 2004-2008. Class I technologies: 0.74% in 2004; 0.983% in 2005; 2.037% in 2006; 2.924% in 2007; and 3.84% in 2008 Solar Electric: 0.01% in 2004; 0.017% in 2005; 0.0393% in 2006; 0.0817% in 2007; and 0.16% in 2008 NJBPU sets requirements for 2009 and after, but must be at or above 2008 levels (see comments regarding proposed RES requirements through 2020)	Class I: solar, wind, geothermal, wave, tidal energy, landfill gas, fuel cells, sustainable biomass Class II: MSW or hydro (<30 MW) that meets high environmental standards
New York	New renewable energy requirement: 0.8% in 2006, increasing ~0.8%/yr to 6.56% in 2013. Customer-sited tier is 2% of total annual RES targets With existing baseline renewable energy, and generation expected from state purchase requirement, renewable energy increases from 19.45% in 2003 to 24% in 2013 (an additional 1% is expected to come from voluntary green pricing programs)	Main Tier: wind, solar, ocean, biomass, biogas, fuel cells, incremental hydro, and low-impact run-of-river hydro > 30 MW Customer Tier: solar, wind (<300kW), fuel cells, and methane digesters
Rhode Island	3% by 2007, increasing 0.5%/yr. to 4.5% in 2010, then increasing by 1 %/yr. to 8.5% in 2014, then increasing by 1.5%/yr. to 16% in 2019 Requirement remains at 16% in 2020 and thereafter unless the PUC determines it is no longer necessary.	Solar, wind, ocean, geothermal, biomass, co-firing, hydro (< 30 MW), fuel cells using renewable resources

There is significant regulatory uncertainty around RPS programs. Of note, the federal Waxman-Markey bill would not preempt the state RPS programs, although it would preempt the state carbon regulation programs between 2012 and 2017. Either a regulatory change in eligible projects or court interpretation of these programs can cause great volatility in REC pricing. As one illustration, Connecticut Class I resources were originally defined to include wind, landfill methane, fuel cell, and solar voltaic resources, and REC prices ranged $35–50 per REC with this definition. However, in June 2003 the state legislature amended the definitions to add certain biomass generation plants located in New England as Class I resources if they reduced NO_x emissions.[11] The Connecticut Department of Public Utility Control (DPUC) made an advisory ruling that an existing biomass plant located in Maine "retooled" to meet a lower NO_x emission standard would qualify to create Class I Connecticut RECs.[12] The market price for Class I RECs came crashing down, dropping the forward price for 2006 RECs by approximately 90%, from the range of $45 down into the vicinity of $3.50. Over time, these prices jumped back to the neighborhood of $30–50/REC.

Legal Concerns Regarding State Programs

Location of renewable resources

States regard the geographic location where RECs are created differently:

- At least three states expressly require that the RECs be created by power generation in the state, and two additional states require that it either be in-state or in the service territory of a state utility [13]—this raises some dormant Commerce Clause issues (see chapter 13).
- Some states require an in-state transmission interconnection to count an out-of-state REC.[14]
- Several states require that RECs actually be associated with energy that is or could be, by virtue of transmission capability that is contracted, delivered in-state.[15]
- Some states allow a wider trading area within an ISO or similar region.[16]

- Some states encourage, but do not require, RECs to be traded in-state by attaching a multiplier value to these in-state RECs.[17]
- Distributed generation typically must be located in the state to qualify to create RECs.

State RPS eligibility rules regarding RECs may limit eligible projects geographically. Some states attempt to limit projects to those constructed within the state or require direct interconnection to the state or state-connected regional grid. Colorado, Illinois, and North Carolina give preferences to in-state projects. Hawaii and Iowa require RPS generation to be from in-state or from the service territory of an in-state utility. California's amendments to its RPS law in 2006 for the first time in a decade allowed new out-of-state generation to be counted toward RPS requirements of load serving entities in the state, removing constitutional issues. Eight states require that the power eligible for RPS RECs must be delivered to in-state load-serving retail power sale entities.

These geographic program restrictions raise Commerce Clause concerns under the U.S. Constitution (see chapter 13). A number of states prohibit the REC credit for out-of-state or out-of-region generation facilities. For example, New England requires that RECs make arrangements on an hourly basis to actually deliver the power to the New England region. New York has a similar system. The NEPOOL Generation Information System (GIS) will only track those resources for RPS credit where out-of-region projects have obtained firm transmission on an hourly basis for sufficient power into the region to equal or exceed the generation from an eligible RPS renewable facility. This does not mean that the exact electrons moved by renewable energy must enter the NEPOOL system. However, enough transmission capacity, in theory, to carry the output of those renewable resources that are deemed to be generating into the NEPOOL region must be at least under contract in order to create credits in a New England state with an RPS program.[18] The largest supply of Massachusetts RECs, about 39%, came from predominantly biomass facilities in the state of Maine, with other New England states providing 17% of RECs and New York and Quebec accounting for 20%. This left only about one-quarter of Massachusetts RECs originating in Massachusetts. Therefore, there is widespread arbitrage in the creation and realization of RPS credits in different states.

Other systems, such as the PJM Generation Attribute Tracking System (GATS), provide a more flexible REC accounting scheme. For the PJM region, this system only requires monthly matching of power from eligible renewable sources out of state to transmission capacity into the

PJM region, in order to qualify for RECs.[19] This longer averaging period is much more accommodating of creating RECs in the PJM region from power generated out of state. This power may or may not be physically moved into the state (but for accounting purposes can show that it could have been moved into the state) over committed transmission capacity over a monthly, rather than an hourly matching period as in NEPOOL.

The PJM interconnect controls power moving over transmission infrastructure in parts or all of 13 Mid-Atlantic states and the District of Columbia. One can only trade RPS credits inside the PJM member states if one is physically located adjacent to the PJM geographic boundary, while certain of the member states, such as Delaware, Maryland, and the District of Columbia, have additional requirements of actual transmission into the system for eligibility. Generators in the New York ISO can trade RECs into Massachusetts, but generators in the PJM control area cannot trade credits into New York.

Legal Ownership of Vintage RECs

The FERC rule on ownership of RECs leaves to states the determination of who owns newly created, and in some cases previous QF-vintage, RECs. Where ownership of RECs is allocated by contract, the contract controls. However, most older QF power sale contracts were silent on this, as they were executed before RECs existed.[20]

The state rulings on this question are split. Some states have granted the RECs to utilities where contracts do not specifically address ownership of the RECs. The majority of these states assigned *vintage* RECs under silent QF contracts to utility purchasers of power; two of the states determined that QFs must be compensated financially for relinquishing title to these vintage RECs, and almost half of the states with *new* contracts allowed RECs to be retained by the power seller where the contract is silent.[21] In the great majority of states that require utilities to net meter power to onsite generators where REC ownership is not explicitly addressed, RECs are allowed to be retained by the net-metered generator.[22] Since traditionally most allowed net-metered projects are limited to relatively small projects, this favors smaller projects over larger ones.

Most states that provide additional cash or other subsidies to renewable generation typically do not address whether any transfer of RECs occurs in return for the funding. However, two states require that in return for funding, any RECs are conveyed to the entity supplying the funding. There does not appear to be a convincing legal rationale as to why exported net-metered power is treated in one manner regarding RECs and stand-alone eligible power generation is treated differently.

Regarding utility implicit ownership of the RECs, there are decisions in Maine where the Maine PUC created an explicit exception to allow purchasers of the entitlements, who do not receive associated GIS certificates, to own the entitlements.[23] Similarly, New Jersey's commission determined that the ownership of RECs would belong to the purchaser.[24] The Connecticut Supreme Court upheld the decision of the Connecticut DPUC, and held that long-term PURPA power sale contracts executed prior to the existence of the Connecticut RECs program did not entitle the project to retain the RECs.[25] To do otherwise, the court reasoned, would provide a windfall, a benefit not bargained for by the power generators at the expense of utility ratepayers.

Other states have taken a pro-generator position. In 2006, California provided that any sale of renewable power prior to 2005 implicitly included the creation of no RECs unless the buyer of power by contract explicitly purchased those RECs.[26] Indeed, under PURPA, federal law, the utility purchaser of power would have been required to enter this contract on these terms, regardless of any state-created RECs.[27] In two matters,[28] the Idaho PUC declined to rule, concluding that Idaho had no "state-created RECs."[29]

RPS Capabilities

Assuming that full compliance is achieved, current mandatory state RPS policies in just those approximately half of the states that have them, will require the addition of roughly 60 GW of new renewable energy capacity by 2025.[30] This amount is equivalent to 4.7% of projected 2025 electricity generation in the United States, and 15% of projected electricity demand growth. The congested and limited state of infrastructure to move renewable power from generation site to market causes some to think that these requirements cannot be achieved within specified time frames (see chapter 12).

Even if states implemented effectively all of their existing current RPS mandates, it would only reduce carbon emissions by 1–1.5% from business-as-usual scenarios by 2015–2020.[31] There is an obvious connection between RPS renewable power programs and goals for carbon reduction strategies: "That RPS are primarily carbon reduction mandates seems relatively clear...this seems to be their primary perceived benefit."[32] RPS renewable power requirements also are not necessarily seen as legally *additional* carbon reductions (see chapter 15). A cap-and-trade carbon reduction program does not guarantee that any renewables whatsoever will be constructed (see chapter 16). However, long-term, electric power is the essential sector for carbon reduction and its investments in power generation are long-term infrastructure realities.

In a number of states, including Nevada, Arizona, New York, and California, new renewable energy project developments are not currently on track to meet mandatory RPS targets for renewable generation as a percentage of total retail load. In some states, extensive exemptions from the RPS purchase mandate, or excuses for retailers not to obtain otherwise required RECs along the lines of legal *force majeure*, have been developed. In several states, regulatory commissions are left broad discretion to grant waivers to regulated entities that do not comply with state RPS requirements. Very open-ended waiver or excuse provisions exist in the RPS programs in Arizona, Hawaii, Minnesota, and Pennsylvania. So the forecast of program implementation of measures and goals is still evolving.

In the next chapter, the focus will shift from developed countries to look at the successful model that has worked to implement renewable power in developing countries. Developing countries are every bit as important—perhaps even more important—than developed countries, over the long term.

Notes

1. Rhode Island and Delaware (partially) have such systems.
2. Such states include Connecticut, Maryland, New Jersey, Washington, D.C., and Texas (partially).
3. Arizona, Colorado, Minnesota, Montana, Nevada, New Jersey, New York, Pennsylvania, and Washington, D.C., are examples of this.
4. Iowa, Massachusetts, Montana (for out-of-state projects), and the Minnesota program covering Xcel Energy are examples of this.
5. California (partially), Colorado, Hawaii, Maine, Minnesota, Montana (for in-state projects), New Mexico, New York (partially), Nevada, Pennsylvania, Texas (partially), and Wisconsin are examples of this.
6. Wiser, R. and G. Barbose. 2008. *Renewable Portfolio Standards in the United States*. Lawrence Berkeley Laboratory. LBNL 154E, (April): 1.

7   Ibid., 1.

8   Ibid.

9   Wiser, Ryan et al. 2007. The Experience with Renewable Portfolio Standards in the United States. *Electricity Journal*. (May): 8, fig. 4.

10 Commonwealth of Massachusetts. Division of Energy Resources. 2006. http://www.mass.gov/Eoeea/docs/doer/rps/rps-2006annual-rpt.pdf.
    MA RPS Annual Compliance Report for 2006. February 15;
    Commonwealth of Massachusetts. Division of Energy Resources. 2007. http://www.mass.gov/Eoeea/docs/dpu/2007_annual_report.pdf.
    MA RPS Annual Compliance Report for 2005. February 20.
    http://www.mass.gov/Eoeea/docs/doer/rps/rps-2005annual-rpt.pdf.

11 See, Connecticut Sub. S. 733 (Public Act 03-135) (2003).

12 See, Connecticut Department of Public Utility Control. 2005. Docket No. 05-03-12, Final Decision (Aug. 10).

13 Iowa, the Xcel Energy requirement in Minnesota, and Hawaii are examples of this.

14 Nevada and Texas are examples of this.

15 Arizona, California, Wisconsin, Minnesota, New Mexico, and New York are examples of this. Delivery can be required on a real-time, monthly, or yearly basis.

16 California, the New England states, Delaware, New Jersey, and Pennsylvania are examples of this, as are multi-jurisdictional utilities. In this case, unbundled RECs can trade apart from the actual energy trade.

17 Colorado, Delaware, and Arizona have attached in-state multipliers to RECs created in the state.

18 See, New York State Public Service Commission. 2006. *Proceeding on Motion of the Commission Regarding a Retail Renewable Portfolio Standard: Order on Delivery Requirements for Imports from Intermittent Generators*. NYPUC Case No. 03-E-0188 (June 28).

19 State of New York Public Service Commission. 2006. Order Authorizing Additional Main Tier Solicitations and Directing Program Modifications. *Proceeding on Motion of the Commission Regarding Retail Renewable Portfolio Standard*. Case 03-E-0188. Issued and effective January 26, 2006.

20 *Wheelabrator Lisbon, Inc. v. Connecticut Department of Public Utility Control*, 531 F.3d 183 [2nd Cir.(Conn.) Jun. 25, 2008]. The case cites American Ref-Fuel Company, Covanta Energy Group, Montenay Power Corporation, and Wheelabrator Technologies Inc., 105 FERC par. 61,004, 2003 WL 22255784 (FERC Oct. 01, 2003)

21 Holt, Ed, Ryan Wiser and Mark Bollinger. 2006. *Who Owns Renewable Energy Certificates? An Exploration of Policy Options and Practice*. Ernest Orlando Lawrence Berkeley National Laboratory. LBNL-59965. As a generalization, vintage QF RECs, where the contract is silent, are vested with the utility purchaser of power, while new contracts, where silent, retain the RECs with the power generator and seller.

22 Ibid. Of the first 12 states to address the issue in the context of net metering, six allowed the generator to retain all RECs. Three states allowed the RECS from onsite use of power to be retained by the generator and the RECs associated with exported excess net power to be vested with the utility (although two of these three required compensation to the customer for that title transfer). One state divided the RECs between the two parties.

23 Maine Public Utilities Commission. 2003. Docket No. 2002-580, 2003 Me. PUC LEXIS 75, February 14.

24 State of New Jersey Board of Public Utilities. 2005. *Transcript of January 12, 2005 Meeting*. (Docket No. E-04080879): 4. As cited in Pennsylvania Public Utility Commission, Doc. No. P-00052149, 35 Pa. Bull. 2041, February 22.

25 *Wheelabrator Lisbon, Inc. v. Connecticut Department of Public Utilities Control*, 531 F.3d 183, 184 (2nd Cir. 2008).

26 California Senate. 2006. Bill 107, chap. 464. http://www.energy.ca.gov/portfolio/documents/sb_107_bill_20060926_chaptered.pdf.

27 *United States Code*, title 16, sec. 824a.

28 Idaho Public Utilities Commission. 2004. *In the Matter of a Petition Filed by Idaho Power Company for an Order Determining Ownership of the Environmental Attributes Associated with a Qualifying Facility Upon Purchase by a Utility of the Energy Produced by a Qualifying Facility*. Case No. IPC-E-04-16; Order No. 29577, 2004 Ida. PUC LEXIS 167, September 3; Idaho Public Utilities Commission. 2004. Case No. IPC-E-04-02; Order No. 29480, 2004 Ida. PUC LEXIS 76, April 27.

29 Federal Energy Regulatory Commission. 2003. Covanta Energy Group, Montenay Power Corporation, and Wheelabrator Technologies, Inc. Docket No. EL03-133-000, 107 FERC par. 61,016, October 1; Re Hearing Denied: American Ref-Fuel Company, Covanta Energy Group, Montenay Power Corporation, Wheelabrator Technologies, Inc., 107 FERC par. 61016, (FERC 2004).

30 Wiser and Barbose, *Renewable Portfolio Standards in the United States*, 1.

31 Engel, Kirsten. 2006. State and Local Climate Change Initiatives: What Is Motivating State and Local Governments to Address a Global Problem and What Does this Say About Federalism and Environmental Law. *Urban Lawyer*. 38: 1,015. Arizona Legal Studies Discussion Paper No. 06-36, www.ssrn.com/abstract=933712.

32 Cabral, Neal. 2007. The Role of Renewable Portfolio Standards in the Context of a National Carbon Cap-and-Trade Program. *Sustainable Development Law & Policy*. American University Law School. 8 (1): 13.

19) THE SUCCESSFUL ARCHITECTURE TO TRANSFORM RENEWABLE POWER

Renewable power can be a long-term win-win scenario for developing nations. Renewable energy can provide opportunities for poverty alleviation, supply power, and enhance energy security by relying on domestic resources. Unlike fossil fuels, renewable resources are widely disseminated across the globe. While many nations—particularly developing nations—have no significant fossil fuel reserves of oil, coal, or natural gas, every nation has significant renewable energy in some form (hydropower, sunlight, wind, agricultural biomass waste, wood, ocean wave power, etc.).

But unless the post-Kyoto architecture segregates and promotes these technologies, developing nations will not deploy them sufficiently, instead opting to burn coal and other traditional fossil fuels. For instance, developers of Kyoto CDM projects in developing nations are trapping methane and flaring it, without turning it into free electricity in the process.[1]

Therefore, while the Kyoto Protocol CDM process encourages carbon reduction in developing countries, it does not always result in a substitution of renewable power for conventional fossil fuel power. The presence of cheap, non-CO_2 credits is a disincentive to developing new CO_2-limiting energy projects that would help to achieve this goal.

CDM architecture must be modified to effect fundamental shifts in energy production. That architecture must target renewable energy projects and forest preservation as priority targets. There will need to be a linkage between national credit and offset trading markets for there to be meaningful international linkage between these markets.

What happens in developing countries is fundamentally important. China currently meets 70% of its electricity demand through coal plants, the most prolific emitters among fossil fuel plants in terms of both CO_2 and particulate matter; 57% of India's electricity comes from coal.[2] While aware of renewable possibilities, each year China adds 40 times more new coal capacity than new wind power capacity. The single-point nature of power plants' emissions, the centralized nature of most power plant decisions in developing nations, and the exploding demand for electricity, make new electricity generating plants the logical frontal assault on GHG emissions.

Of the countries with the largest coal reserves—United States, China, India, or Indonesia—none has a carbon policy to regulate the release of CO_2 from the deployment of such coal reserves. India, in fact, in the latter part of 2008, came looking to acquire ownership of existing coal mines in the United States and elsewhere, to fuel its coal-fired power industry.[3] Looking to spend $4 billion, it was stated that for coal resources, "Money is not a problem."

The funding now of renewable energy projects worldwide, and especially in Asia given its continuing rapid growth and industrialization, is necessary to prevent these nations from becoming even more reliant long term on an expanding fossil fuel-based generation infrastructure. There needs to be a proven format that developing nations can deploy to achieve the right kind of renewable energy power projects. Fortunately, a proven model for the new post-Kyoto architecture exists. It is detailed in the book by Steven Ferrey and Anil Cabraal, *Renewable Power in Developing Countries: Winning the War on Global Warming*, published by PennWell. While key factors are described in the following sections, readers are referred to this recent separate book, which contains detailed discussion not only of these successful programs and model, but of environmental impact assessments, financing mechanisms, the role of international agencies, and various applicable laws that determine the legal relationship of stakeholders.

The Successful Power Sector Model in Developing Nations

There is hard evidence of the model for success in developing nations deploying renewable power.[4] For the past decade, since the original decision to implement international carbon controls, five nations in Asia were among the first in developing small power producer (SPP) programs to promote renewable energy development in their countries. These programs create an important model of best practices for the new architecture for post-Kyoto investment.

Some of these have achieved in just a few years a substantial contribution from new renewable small power projects to their growing national energy supplies: almost 4% of power supply in Sri Lanka, India, and Thailand are from small power producer (SPP) independent renewable energy initiatives[5] This is the appropriate laboratory, as approximately 60% of all new power generation capacity financed in developing countries is in Asia. These nations feature different forms of government and have different predominant fuel sources in their power generation bases (hydro, coal, gas, and oil). Some of the national electric systems have an integrated high-voltage transmission system, whereas others have a disintegrated or island system. These successes in Asia are the architecture for developing countries' participation in the post-Kyoto scheme.

Table 19-1 displays key comparative elements of program design and implementation in these five non-Annex I Asian programs. The middle column illustrates that two of the five profiled programs subsidize renewable energy SPPs. Thailand does so by providing a project-specific subsidy through a competitive solicitation process. Andhra Pradesh state in India does so by providing a feed-in tariff in excess of true avoided cost for renewable energy SPP power sales.

Table 19–1. Comparative developing nation renewable program overview

Country Program	Year begun	Maximum size (MW)	Premium for renewable energy	Primary fuel	Eligible PPA solicitation
Thailand	1992	<60 or <90	Yes, competitive bid	Gas	Controlled period
Indonesia	1993	<30 Java <15 other island grids	No	Energy	Controlled period
Sri Lanka	1998	<10	No	Hydro	Open offer
India: Andhra Pradesh	1995	<20 Prior <50	Yes, in tariff	Wind	Open offer
India: Tamil Nadu	1995	< 50	No	Wind	Open offer

Several important lessons for small renewable program design and policy in developing nations are revealed by analyzing these programs in detail. A national or international framework for structured SPP project development is necessary: SPPs will not spring full-borne from the existing electric sector environments in developing countries. A system of law, regulation, and utility interface that facilitates orderly SPP renewable power development must exist. Many of the programs were implemented prior to the Kyoto CDM program, which provides even greater monetary incentives. There is a model for successfully shifting to a renewable power base in fast-electrifying developing nations.

The following sections give a quick overview of some of these developing country programs.

Sri Lanka

The experience in Sri Lanka is indicative that successful renewable energy projects can be implemented even under the most difficult circumstances. From the mid-1980s until the spring of 2009, Sri Lanka was engaged in a protracted civil war with the Tamil Tigers, an ethnic Tamil minority within the Sinhalese-dominated country. This bloody succession movement claimed about 80,000 lives, including a former sitting prime minister of the nation.

Sri Lanka is a single island nation. A single national electric power utility, the Ceylon Electricity Board (CEB), maintains a monopoly on retail power sales. As of 2007, the national utility grid in Sri Lanka had 1,800 MW of installed generation, double from a decade earlier. A single standardized power purchase agreement (PPA) and standardized power purchase tariff, designed to be fair to both private small renewable power producers and the purchasing utility, were the foundation for this renewable energy program.

Fifteen-year PPAs are available for projects up to 10 MW in size. This threshold is now increased to 20 MW as the maximum eligible size for renewable energy projects. To attract wind and biomass projects, Sri Lanka moved to a cost-based PPA feed-in tariff for SPPs that is differentiated for each renewable technology, so that wind and biomass will receive a higher tariff than small hydro projects (see chapter 17).[6]

Small hydro and other renewable energy developers of facilities no larger than this threshold are allowed to sign a standardized PPA with CEB. All of the awarded projects were small hydroelectric projects with the single exception of one small cogeneration facility. As of 2007, Sri Lanka had 52 operating SPPs supplying more than 100 MW of power.

Another 25 SPPs were under construction, plus another approximately 25 SPPs under active development. The term of the PPA is up to 15 years, and because of program success, expanded to 20 years as of 2007.

Table 19-2 sets forth in abbreviated format principal elements of the Sri Lanka SPP program, including a peak-season tariff differentiation and a rolling SPP award process.

Table 19–2. Primary elements of the Sri Lanka SPP program

1. Process: Open offer
2. Maximum size: 10 MW
3. Tariff: Avoided cost for nondispatchable projects de facto capped not to exceed tariff paid to larger IPPs. Differentiated for wet and dry seasons.   Wet season: SL Rs. 5.85 per kWh [$0.06]   Dry season: SL Rs. 6.06 per kWh [$0.062] (2003)
4. Third-party retail sales: No
5. Self-wheeling: No
6. Energy banking: No
7. Standardized PPA: Yes
8. PPA term: <15 years
9. Subsidy or incentives: SPP and IPP power equipment generally exempt from import tax and enjoy tax holiday if projects are implemented under Board of Investment rules (http://www.boi.lk).

Thailand

Thailand was one of the first countries in Asia to adopt a small power solicitation program. The Electricity Generation Authority of Thailand (EGAT) has installed about 22,000 MW of generation capacity. The Thai program was modeled on elements of the PURPA small power program in the United States.

What is of particular note in the Thai system is that competitive bidding by renewable energy SPPs is used to suppress and award subsidy payments. It has been successful in minimizing the cost of such subsidies and employing available subsidy funds to bring forth the maximum number of megawatts of new private power resources. However, such a competitive system requires that there be a controlled competitive solicitation process for SPPs. By contrast, India's states and Sri Lanka avoid a solicitation in favor of a continually open offer to sign PPAs and purchase power.

A first request of 300 MW was made for SPP power by EGAT in 1992. This amount was expanded to 1,444 MW in late 1995. Eligible projects include biomass, waste, mini-hydro projects, photovoltaic (PV) systems, or other renewable projects, such as wind.

The regulations allow SPPs to deliver for sale to EGAT up to 60 MW, although up to 90 MW is within the discretion of EGAT to accept on a case-by-case basis. Several projects at 90 MW have been accepted. Of note, most of the Thai SPP projects are not renewable projects, but are gas-fired. Although the program is a success and has demonstrated SPP cogeneration potential, it has not restricted participation to renewable sources.[7]

Table 19-3 sets forth in abbreviated format the primary program design, tariff, and contract provisions of the Thai SPP program, including its innovative renewable subsidy incentive.

Table 19-3. Primary elements of the Thai SPP program

1. Process: Controlled solicitation
2. Maximum size: 60 MW (90 MW with permission)
3. Tariff: Avoided cost to utility
   For firm 20 year energy and capacity:
   Coal: $0.04 per kWh
   Gas and Renewables: 2.14 baht per kWh assuming 85% capacity factor [$0.051 per kWh (2003 exchange rates)]
4. Third-party retail sales: No. Under consideration
5. Self-wheeling: No. Under consideration
6. Energy banking: Only for SPPs <1 MW
7. Standardized PPA: Yes. After 2001, because of excess capacity, EGAT purchases 100% of capacity rating of kWh on peak and 65% of capacity rating kWh off-peak. Therefore, project cannot supply and be paid for rated capability during off-peak periods.
8. PPA term: Firm, 5-25 years
   Nonfirm, <5 years
9. Subsidy or incentives: Competitive bidding for five-year renewable subsidy.
   Up to $0.009 per kWh based on lowest bids.
   Eight-year income tax holiday.
   Equipment exempt from import tax.

Even as of the end of 2002, 71 SPPs had been accepted and obligated, with a total capacity of 2,330 MW. Thirty-five of these SPPs for 2,048 MW were firm commitments, whereas 36 projects for 282 MW were shorter-term non-firm commitments. The bulk of these projects are cogeneration projects, and most of these firm power projects are powered by natural gas. It is also typical of these renewable SPP projects that they are cogenerating power for self-use and exporting less than the installed capacity. Many of those have an installed capacity above 90 MW and contracted to sell EGAT 90 MW. Contract terms of 20-25 years are the norm for these larger cogeneration projects under firm contracts.

The renewable energy projects are primarily fuel by rice husks (sometimes augmented by wood chips) and bagasse (sugar mill waste).

Also represented are projects fired by wood waste and solid waste. Many of the renewable SPPs are much smaller (1–8 MW) and do not have long-term contracts, or they have non-firm contracts that are extendable by EGAT. A few of the renewable SPPs have set 5- or 10-year contracts.

India

In India, about 42–44% of the rural population has access to electricity.[8] India has become a major player in renewable generation and private sector power development. India is the world's 10th largest developer of small hydro facilities, and the 5th largest developer of wind power, as well as the 5th largest producer of PV systems, in the world. In India, state electricity boards provide electric power. Much of the authority for electricity policy resides at the state, rather than federal, level. A number of states have SPP programs.

The Ministry of Non-Conventional Energy Sources (MNES) at the federal level promotes renewable power through grants and subsidies, so as to create a level playing field for various energy sources. Certain renewable energy technologies also receive preferential federal tax treatment.[9] Table 19-4 scales the realization of various renewable energy development in India against its potential. More than 3,400 MW of renewable projects were in operation by the end of 2001, from a base of about 100 MW in 1992 before these initiatives began.[10]

Table 19-4. Realization of renewable projects in India, as of 2003

Technology	Potential (MW)	Realization (MW)
Wind	45,000	1,267
Small hydro up to 25 MW	15,000	1,341
Biomass power	19,500	308
Biomass cogeneration	3,500	273
Urban and industrial waste	1,700	15
Photovoltaics	Significant	47

Andhra Pradesh, India. Of the more than 30 Indian states, the state of Andhra Pradesh is the most advanced in installing wind capacity.[12] Andhra Pradesh has a system of more than 7,000 MW that is short of capacity to serve existing demand. In Andhra Pradesh, there was no formal solicitation or bidding; project developers could submit requests that were independently judged. Applications for a contract can be submitted at any time on a standardized form.

The transmission utility, Transmission Corporation of Andhra Pradesh Limited (APTransco), which is distinct from the state electricity board—APSEB, the distributor—unilaterally makes the initial decision on whether to enter into a PPA with an SPP developer. This same entity that makes this decision selects the rate it is willing to pay to purchase the power from such a facility, and negotiates a PPA with the project. SPP project eligibility was originally for projects up to 50 MW, and has since been lowered to a 20 MW maximum. Any renewable or nonconventional waste technology is eligible. Andhra Pradesh has approved the construction of 1,013 MW of nonconventional generation. This is scaled against the potential in table 19-5.

Table 19-5. Andhra Pradesh renewable project status, 2003

Technology SPP	Projects approved	Projects complete	Projects at finance close	Potential capacity
Wind	283	9?	10	745
Biomass	345	81.5	110.7	627
Bagasse cogeneration	210	49.5	75.5	250
Municipal waste	23.6	0	0	40
Industrial waste	36	1.5	4	135
Small hydroelectric	95	69	30.4	1,252

Capacity (MW)

By 2005, 189 MW of capacity were in operation; the remainder is in various stages of development. There is no formal standardized contract. Therefore, individual negotiation occurs with the state utility monopoly to determine the contract terms. Table 19-6 sets forth in abbreviated format relevant SPP provisions of the SPP program and tariff in Andhra Pradesh, including the significant wheeling fee.

Table 19-6. Primary elements of the Andhra Pradesh SPP program

1. Process: Open offer
2. Maximum size: < 20 MW (was < 50 MW)
3. Tariff: Above avoided cost to utility not to exceed 90% of industrial retail tariff
    Rs. 3.32 per kWh ($0.0698 per kWh at 2003 rates)
4. Third-party retail sales: No (previously allowed)
5. Self-wheeling: Allowed with 28% wheeling fee plus $0.01 per kWh charge
6. Energy banking: Allowed with 2% energy banking charge
7. Standardized PPA: Yes
8. PPA term: 20 years
9. Subsidy or incentives: Federal loans with 1- to 3-year repayment moratorium
    80% of capital cost can be depreciated against taxes in the first year
    Grants for PV systems
    Equipment exempt from sales tax

The state utility makes the determination of the purchase rate it will offer the SPP.[12] The utility has attempted to abrogate existing PPA tariffs. Aggrieved SPPs have appealed the rate to the Andhra Pradesh Electricity Regulatory Commission (APERC), which has set the rate on appeal. Decisions of APERC have also been appealed by aggrieved SPPs to the Supreme Court. The utility would prefer not to purchase power from SPPs, and APTransco has staked out a position opposed to certain expansion of renewable energy SPPs.[13]

Tamil Nadu, India. The Tamil Nadu system has more than 7,000 MW capacity. About 86% of villages, but less than half of households, are electrified. Tamil Nadu state has a significant fraction of India's wind turbine capacity and a significant percentage of biomass projects.

An SPP size limit of 50 MW is imposed (table 19–7). Multiple generator sets at the same location can be separately packaged into separate applications to effectively exceed this limit at the site. Also, if the power is wheeled for one's own consumption at a remote location, rather than sold to the grid, more than 50 MW is allowed.

In Tamil Nadu state, no formal standardized PPA is employed, although the utility has employed the same PPA of its design in every situation, thereby creating a de facto standardized PPA. This again leaves great discretion with the utility. Wheeling of power to an affiliated location—not to a third-party—is permitted. The tariff is higher for biomass projects than for wind, to reflect the former's non-intermittent, controllable power generation characteristics. Table 19–7 sets forth in abbreviated format salient elements of the Tamil Nadu SPP program, including its low wheeling charge, especially compared to that of its neighbor state of Andhra Pradesh.

Table 19–7. Primary elements of the Tamil Nadu SPP program

1. Process: Open offer
2. Maximum size: < 50 MW
3. Tariff: Above avoided cost to utility not to exceed 90% of industrial retail tariff
   Wind: Rs. 2.7 per kWh ($0.057 per kWh at 2003 rates)
   Biomass: Rs. 2.88 per kWh ($0.06 per kWh at 2003)
4. Third-party retail sales: No (previously allowed)
5. Self-wheeling: Allowed with 2% wheeling charge for up to 25 km transmission;
   10% wheeling charge more than 25 km
6. Energy banking: Allowed with 2% banking charge
7. Standardized PPA: Yes, in final development
8. PPA term: 5–15 years
9. Subsidy or incentives: 80% of capital cost can be depreciated against taxes in the first year
   Grants for PV systems
   Equipment exempt from sales tax

Most of the SPP projects are wind, bagasse, cogeneration, biomass gasifier, and photovoltaic. Tamil Nadu is a major locus of wind power generation. There is a wind turbine testing laboratory in the state. In Tamil Nadu, about 80% of wind-generated power is used captively by the owners, and about 20% sold to the state utility, TNEB. As of September 2002, the awards displayed in table 19-8 were in place.

Table 19-8. Tamil Nadu awards made

Energy source	Output (MW)
Wind	894
Bagasse	186
Other biomass	13
Photovoltaic (PV)	2

Solving the International Carbon Equation

The choices of technology are critical. With population increasing precipitously, it is the policy of nations and international organizations to reduce poverty and increase standard of living, both of which are directly correlated with electrification.[14] With increased standard of living and development, inexorably comes increased consumption of energy. No national governments are encouraging their people to forsake the conveniences of development, even to help abate global climate disaster. Exponential increases in electric demand and production in developing nations are unavoidable, both as a natural progression of modern societies, but also as a development goal encouraged by national governments and international agencies.

The choice of energy technology is one of the few inputs in the global warming algorithm that can be influenced significantly by law and policy. If developing nations do not direct the choice of renewable power options in lieu of conventional reliance on fossil fuel power technologies, the war on global warming cannot be won. Otherwise, the business-as-usual increase in electrification in developing countries will totally swamp the cuts that would be accomplished if the Kyoto Protocol were to achieve its targets in the developed nations that it affects.

Many developing countries do not, alone, have access to sufficient resources to expand their electric infrastructure. Electric power is the most capital-intensive industry in the world by far.[15] Renewable power resources are even more capital intensive than conventional power resources, and thus require even larger commitments of capital to initiate. But once constructed, they offer the advantage of not requiring expensive fuel sources to continue operation. Therefore, renewable power generation, if it can be financed, can actually be an asset for a developing country.

Ultimately, the challenge is not technological, nor even financial. The challenge is legal and regulatory: the missing link is the institutional mechanism and model to steer and implement low-carbon power choices in developing countries. When one analyzes some of the early efforts to combat global warming emissions in developing nations, there emerges a model of best practices for how to structure a low-carbon high development growth curve for developing nations.

In developing nations, where the growth of electric power requirements and potential emission of greenhouse gases are most extreme, there seldom is a competitive deregulated power market, as characterizes much of the U.S. power market.[16] Therefore, this centralization of legal decision making affords the ability to implement a more sustainable global warming strategy in developing nations. A legal and regulatory model now is emerging from analysis of early efforts that yield successful renewable low-carbon power development programs in developing nations.

To accomplish this, there needs to be a fair and equitable long-term power purchase agreement that memorializes this relationship.[17] The length of the term of the PPA must be at least as long as the contemplated loan to the developer for construction of a capital-intensive renewable power project, in order to secure financing. To create such a neutral and fair PPA, the successful model used by those developing nations dictates that international legal consults/attorneys be retained to meet with stakeholders and carefully devise a PPA that meets the objectives of the program.[18] These attorneys also need to enlist in the PPA a fair dispute resolution mechanism.[19] Within the framework of a fair and neutral PPA, in which is contained an avoided cost-based tariff, the various choices of successful program design and best practices can be molded into a successful renewable power program. There are several important elements:

Transparent regulatory process. A transparent process is required to build investor, developer, and lender confidence.

Standardized PPA. All programs employ either de jure or de facto standardized PPAs, and most employ either an avoided cost-based tariff or avoided cost principles. All allow some form of long-term firm contract commitment.

Legal dispute resolution mechanism. A legal framework for structured project development is necessary.

Allocation of legal risks. A variety of commercial, sovereign, currency, and regulatory risks are implicitly or expressly allocated in the power sector. The Thai program reduces the SPP capacity payment where the SPP does not deliver. Tamil Nadu facilitates SPP power wheeling.

Interconnection requirements. Utilities must interconnect with renewable energy projects subject to a straightforward procedure to accomplish this without significant transaction costs or interconnection risk.

Legal milestones and bid security. To eliminate the speculative non-development risk, the Thai program requires a bid security deposit of 500 baht per kW ($12 per kW) of capacity. Sri Lanka in 2003 placed a six-month limit on the validity of letters of intent (LoIs) granted for renewable projects and required bid security bonds of SLRs, 2,000 per kW ($20 per kW). The Thai program also requires a deposit of 100 baht per kW ($2.50 per kW) of applicants, and more for larger sources.

Renewable set-aside. The program in Thailand allocates entitlements and subsidies in order of the most preferred projects and the least required subsidy for renewable projects. A variant in 28 U.S. states employs a renewable portfolio standard for a minimum percentage of power sold by each retail seller (see chapter 18).

Third-party sales. None of these Asian programs currently allows direct third-party retail sales of power by the SPP (except in limited industrial estate areas). However, other states in India do allow direct retail sales.

Net metering and energy banking. Energy banking is allowed in more than half the states in the United States in the form of *net metering*, providing an exchange of renewable energy during a billing period (see chapter 17). Several of the Asian countries adopted energy banking variants, and in 2009, Sri Lanka adopted net metering.

The next chapter turns to a natural biological mechanism to sequester carbon, through living forests. These are renewable in every embodiment of that concept. Preservation or reconstruction of these resources is an important policy consideration in carbon control mechanisms.

Notes

1. Observation of the author in his work for the World Bank and UN Development Program as legal advisor on power development in developing countries since 1993 to the present. See also, Christen, Kris. 2004. Environmental Impacts of Gas Flaring, Venting Add Up. *Environmental Science and Technology*. Nnimmo Bassey. Gas Flaring: Assaulting Communities, Jeopardizing the World. Dec. 2008, available at http://www.eraction.org/publications/presentations/gas-flaring-ncc-abuja.pdf; John Uhren & Joseph Doucet. *The Flaring of Associated Gas*. Dec. 15, 2004, available at http://apps.business.ualberta.ca/cabree/pdf/2004_Fall_Projects/John%20Uhren.pdf. GAO: Natural Resources and Environment. 2004. *Natural Gas Flaring and Venting: Opportunities to Improve Data and Reduce Emissions*. www.gao.gov/cgi-bin/getrpt?GAO-04-809.

2. French, Howard. 2004. China's Boom Brings Fear of an Electricity Breakdown. *New York Times*. July 5: A4.

3. Zeller, Tom. 2008. India Shopping for Coal Mines in Appalachia. *New York Times*. October 23.

4. See, Ferrey, Steven and Anil Cabraal. 2006. *Renewable Power in Developing Countries: Winning the War on Global Warming*. Tulsa: PennWell. Chap. 9.

5. For statistics on the Thai and Sri Lanka programs, see EGAT, available at http://egat.co.th/en; and Ceylon Electricity Board, available at http://www.ceb.lk, respectively. Alternatively, http://www.boi.lk lists incentive investment details for Sri Lanka. A review of programs, tender notices, and a series of reports on the India program is available at www.MNES.nic.in.

6. See, CEB Web site, at http://www.ceb.lk/.

7. Subsidies are available in the 2001–2002 solicitation process for up to five years for renewable projects in the amount of not more than 0.36 baht per kWh ($0.01 per kWh). The subsidies are granted under the Energy Conservation Promotion Fund Committee (ENCON), established by the Energy Conservation Promotion Act, B.E. 2535 (1992). Two billion baht ($50 million) is allocated to such renewable project subsidies, in up to 300 MW of such projects contracted after June 2000.

8. Ministry of Power. 2003. *Discussion Paper on Rural Electrification Policies*. New Delhi: Government of India, (November): 8. This leaves almost 80 million rural households without access to electricity. Notwithstanding this statistic, as of March 2003, 87% of inhabited Indian villages were declared electrified.

9. Ministry of Non-Conventional Energy Sources. 2001. *Renewable Energy in India*. New Delhi: Government of India. 53ff, 57. Wind energy components pay a much lesser customs import duty than assembled wind generators, whereas PV systems are exempt from excise duty. Accellerated depreciation is allowed for solar devices, biogas equipment, wind turbines, and agricultural and municipal waste conversion equipment.

10 Ministry of Non-Conventional Energy Sources. 2002. *Annual Report 2001–2002*. New Delhi: Government of India. 7; from confidential World Bank data and reports, 2002. This increased the percentage of renewable energy in India from 0.13% in 1992 to 3.4% in 2001. Wind accounts for almost half of this renewable capacity; 26 interconnected PV projects are in service. In addition, there are 400 sugar mills where renewable energy cogeneration is possible, as well as a potential of 16,000 additional biomass opportunities. There is estimated to be the potential for 15,000 MW of small hydro potential (of 25 MW each of larger).

11 MNES, *Renewable Energy in India*, 8.

12 The Andhra Pradesh Electricity Regulatory Commission (APERC) issued G.O.MS No. 93 of 1997, and G.O.MS No. 112 of 1998, as extended, which provide the rates paid to non-conventional energy projects. The 1997 order is designed to encourage renewable energy resources, by establishing a power purchase price of Rs. 2.25 per kWh, escalating at 5% per year. It also allows third-party sales and energy banking. However, the 1998 order clarifies that, in the absence of the grant of a license pursuant to the Electricity Duty Act of 1939, SPPs cannot make third-party sales.

13 Personal communication (2002) in Hyderabad, Andhra Pradesh. It claimed that its marginal cost of new combined-cycle gas-fired power to be about Rs.1.8 per kWh ($0.035 per kWh). The utility also is concerned that they cannot control where SPPs choose to site projects.

14 See the About Us and Challenges pages at www.worldbank.org.

15 International Energy Agency. 2003. *World Energy Investment Outlook: 2003 Insights.* 65, 345.

16 See, generally, Ferrey, Steven. 2000. *The New Rules: A Guide to Electric Market Regulation*. Tulsa: PennWell. Chap. 77.

17 For a discussion of elements that go into the creation of a PPA, see Ferrey, Steven. 2010. *The Law of Independent Power*. Eagan, Minnesota: Thomson/West Publishing. Sec. 3:77.

18 The author has performed this role in several developing countries on behalf of the World Bank and the United Nations Development Programme.

19 In the author's experience, this role typically might be played by a utility regulatory authority, such as the Federal Energy Regulatory Commission and state public utility commissions in the United States. However, where there is not a competitive market and/or the utility is a state-affiliated entity that reports to a government ministry in a developing country, there may not be a utility regulatory authority. Creating such entities is part of electric sector reform pushed by many international agencies. However, once established, it requires years, rather than months, for these new authorities to build capabilities, hire staff, issue regulations, and establish themselves as an effective dispute resolution mechanism.

20 INTO THE WOODS

"We are recklessly burning and clearing our forests and driving more and more species into extinction. The very web of life on which we depend is being ripped and frayed."[1]

The Role of Forests in Biologically Converting Carbon Accumulation

What is renewable, natural, self-sustaining, and given no credit under any current international or national climate change programs? It might be surprising to learn what the answer is: existing forests. Developing countries harbor some of the great forests of the world. The great rain forests of South America and Indonesia are primary testament. Notwithstanding the lack of regulatory recognition, forests matter.

Forests are renewable resources. But it is not their potential role as a harvested renewable biomass power generation fuel that makes them valuable in addressing global warming or eligible to create carbon offsets. It is the role of living forests in naturally absorbing and converting carbon molecules in their natural life cycles that makes them eligible for offset creation and credit. The net concentration of CO_2 in the environment is a function not only of the output of CO_2, but the conversion of CO_2. Plant life and the oceans are the great biological carbon sponges, absorbing and converting CO_2. Forests are an opportunity to naturally sequester CO_2

in the atmosphere, rather than the proposed controversial sequestration experiments proposed underground in mines or in the ocean. The agricultural sector offers a significant potential for sequestration of carbon through preservation of biologic resources.

Forests use carbon dioxide as building blocks for organic molecules and store it in woody tissues, but that process is not indefinite. The United States estimates that in 2006 it had 7,202 million metric tons of carbon dioxide-equivalent emissions, but that carbon sinks absorbed or converted approximately 900 million metric tons of these carbon dioxide equivalents, reducing the total by about 15% to 6,319 net million metric tons of emissions.[2] The carbon stored in the existing forests of the contiguous 48 U.S. states equals about 20 years of industrial U.S. carbon emissions.[3]

Preservation of existing forests would revolve around preventing deforestation. Deforestation is the conversion of forested land typically for arable land for urban use, timber, or wasteland. More than half of the naturally forested areas of the world have been destroyed or damaged during the past 8,000 years, with half of that occurring in the past 50 years. Rain forests in the Brazilian Amazon, Madagascar, and the Philippine archipelago are seeing soaring rates of deforestation.[4] Thus the rate of deforestation has accelerated dramatically. About half of the mature tropical forests, between 750 to 800 million hectares of the original 1.5 to 1.6 billion hectares that once covered the planet, have already fallen.

Six billion tons of carbon are released each year to the environment due to deforestation. Annually in the world, about 32 million acres (13 million hectares) of forest are destroyed and not replanted.[5] Forests play a critical role in carbon concentrations. Yet they face destruction on an unprecedented scale as local populations seek additional land for farming and ranching, and they provide a source for lumber for commercial interests.

Deforestation accounts for 18–25% of global annual carbon emissions. Annually, there are depletions of forests equal to the size of Portugal, or half the size of England.[6] Half of this destruction is from illegal logging that is not effectively policed.[7] About 70% of the wood cut in Indonesia is illegal.[8] This costs the Indonesian government about $1 billion annually in lost taxes, royalties, and lease payments. Collectively, these emissions from two countries account for almost 10% of the world's total emissions of greenhouse gases. A World Bank report found that 83% of Indonesia's annual greenhouse gas emissions, as well as 60% of Brazil's greenhouse gas emissions, come from the destruction of their forests.[9]

Yet, news from Indonesia features not its rainforest protection, but a crash program to build a significant number of new coal-fired power plants without sequestration of carbon emissions.[10] In March 2008, Indonesia invited banks to participate in financing five new coal-fired power plants worth more than $2 billion, representing the first part of an effort by the world's fourth most populous country to almost double their generating capacity utilizing coal-fired generation. This is an important loss: forest preservation is most valuable in the tropics, as forests in snowy climates can retard the reflection of warmth by snow cover.[11] Tropical forests are responsible for about one-quarter of carbon absorption, and this is where deforestation is fastest: tropical forests in South America and Africa.

Land use and land-use change and forestry (LULUCF) describes the terrestrial means to accumulate carbon in the ecosystem. Changes in percentage emissions by country, including LULUCF, during the period 1990–2007, is charted in figure 20–1. However, it is not uniformly valued as a carbon credit. Legally there has emerged a difference between preserving existing forest, versus creating or afforesting new areas. For forestation projects, the integrity, additionality, and credibility of offsets credited to new afforestation also have become controversial issues (see chapter 15). Forestation creates an opportunity to improve livelihoods in rural areas of developing countries.

International Carbon Regulation and Forest Credits

This section looks at how international protocols treat forests. Leading carbon scientists have submitted that the only way to reduce carbon concentrations to even 90% of current levels is to adopt "forestry practices that sequester carbon" or there will be "irreversible catastrophic effects."[12] Notwithstanding this biologic role in converting carbon, the Kyoto Protocol does not recognize the preservation of forest for creating creditable offsets that can be traded to those needing carbon emission allowances.

280　UNLOCKING THE GLOBAL WARMING TOOLBOX

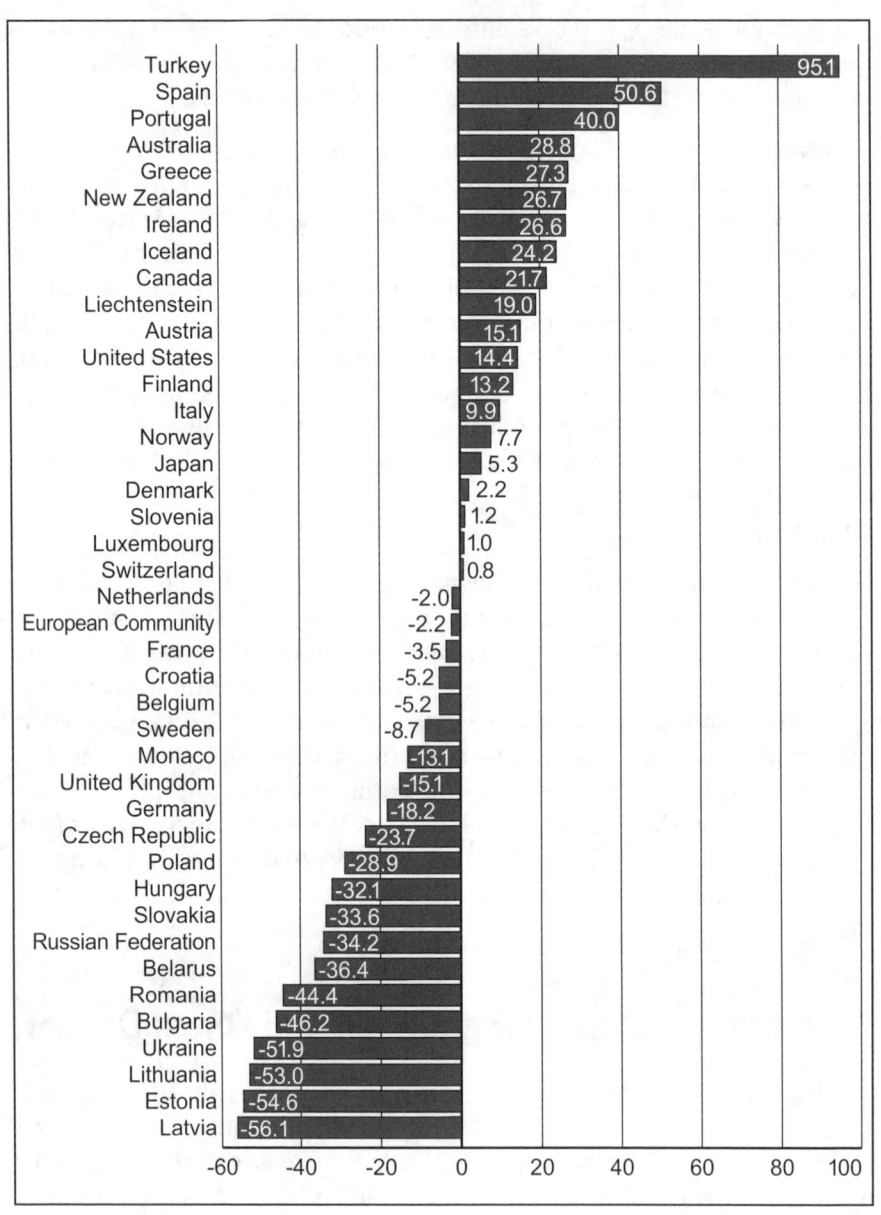

Country	%
Turkey	95.1
Spain	50.6
Portugal	40.0
Australia	28.8
Greece	27.3
New Zealand	26.7
Ireland	26.6
Iceland	24.2
Canada	21.7
Liechtenstein	19.0
Austria	15.1
United States	14.4
Finland	13.2
Italy	9.9
Norway	7.7
Japan	5.3
Denmark	2.2
Slovenia	1.2
Luxembourg	1.0
Switzerland	0.8
Netherlands	-2.0
European Community	-2.2
France	-3.5
Croatia	-5.2
Belgium	-5.2
Sweden	-8.7
Monaco	-13.1
United Kingdom	-15.1
Germany	-18.2
Czech Republic	-23.7
Poland	-28.9
Hungary	-32.1
Slovakia	-33.6
Russian Federation	-34.2
Belarus	-36.4
Romania	-44.4
Bulgaria	-46.2
Ukraine	-51.9
Lithuania	-53.0
Estonia	-54.6
Latvia	-56.1

Fig. 20-1. Changes in percentage GHG emissions by country, including LULUCF 1990-2007

The Kyoto Protocol does not cover developing countries, and in some ways even fails to speak their language. Under the Kyoto Protocol, offset carbon credits can be obtained for planting trees but not for preserving existing forests. Leading up to the Kyoto Protocol, developed nations objected to such credit, arguing that it would be difficult to monitor and measure the amount actually preserved, as well as to ensure that preservation would endure over time. It is estimated that purchasing and policing existing forest preservation could be accomplished at a price of approximately $4 per ton of carbon saved, much cheaper than the cost of EU carbon allowances or credits. No credit of any kind either under the Kyoto Protocol or under the parallel European Union Emission Trading Scheme (EU-ETS) is given to any countries for preserving forest resources. Some modest demonstration programs on preserving forests were finally launched in 2008.

The Kyoto signatories expressly limited land use and forestry projects that can earn credits to those involving afforestation (tree planting on non-forested land) and reforestation (tree planting on previously forested land).[13] The EU-ETS, which provides the chief Kyoto compliance mechanism for the union's members, goes even further, excluding all forestry credits, including those from afforestation and reforestation projects.

Much of the current debate on forest protection focuses on who controls the money: the developed country donors of the funds, intermediary international funding agencies, or the indigenous persons who reside in the areas where either forest is desired to be protected or carbon-emissions capped. Some have criticized the World Bank's substantial efforts to fund forest preservation as benefiting private interests who preserve forest more than they preserve locally affected communities. There were protests in late 2007 by developing countries and their citizens against the establishment of the World Bank funded Forest Carbon Participation Project for forest preservation projects. So while this is about carbon, it is also about control over decisions about world resources and projects.

Efforts of some developing countries to change the protocol to include for credit the act of avoiding deforestation in a CDM project was tabled in 2005 meetings, and not resolved at the Bali Kyoto meetings in late 2007.[14] Thus, conservation of forests has been tabled until after 2012 by the Kyoto parties. The UN IPCC report notes that forest offsets under Kyoto "is being lost in the current institutional context and lack of political will to implement and has resulted in only a small portion of this potential being realized at present."[15] At its affordable cost, the most cost-effective solution in developing countries for carbon concentration mitigation is off the Kyoto table and needs to be resurrected as part of creating a new architectural line of cost-effective activities.

Eligibility: New Forest; Old Forest

In the 1960s, a study (using 10 years worth of data from a single plantation) suggested that forests 150 or more years old give off as much carbon as they take up from the atmosphere, and are thus carbon neutral.[15] More recent current data suggests that carbon accumulation can continue in forests indefinitely.[16] When an old growth forest is harvested, there's a new input of carbon to the atmosphere for about 5–20 years, before the growing young trees begin to absorb and sequester more carbon than they give off.

However, these old growth forests around the world are not protected by international treaties and have been considered of no significance in national "carbon budgets" under the Kyoto Protocol. They do not create an offset under Kyoto, EU-ETS, or RGGI. There is a significant carbon sequestration loss once original natural forest is destroyed, even if replaced with new afforestation acreage that can earn offsets. However, preservation of forests is not eligible to create any credits under either the Kyoto Protocol or RGGI.

Apparently, though, not all forest resources are created equal: there is new evidence that afforestation, recognized under the Kyoto Protocol for offset credits, may not be as beneficial as forest preservation, not recognized under the Kyoto Protocol.[17] An Australian study claims that natural forests are 60% more efficient than created forests in sequestering carbon. So perhaps preservation of existing forests is more critical than afforestation of new areas, which are eligible. If so, there is a significant carbon sequestration loss once original natural forest is destroyed, even if replaced with new afforestation acreage. However, preservation of forests is not eligible to create any credits under the Kyoto Protocol, the EU-ETS, or the U.S. RGGI program now operating in its 10 states.

U.S. Carbon Legislation and Forests

Forest issues have been significant in the last several years in U.S. carbon legislation.

Forest eligibility has been controversial in the consideration of U.S. federal legislation. In 2008 there was an effort to expand the farming and forestry offsets available under the 2008 Lieberman-Warner carbon legislation proposal, which included no-till agricultural practices. It was amended to allow a larger amount of international forest projects to create offsets, allowing up to 15% use of all offsets to create compliance. However, bipartisan-supported amendments in the Senate sought to eliminate offsets created overseas, as that would encourage carbon spending in developing countries instead of in the United States. Various U.S. congressional bills would include a provision to cause imported goods from countries lacking carbon controls to be covered by emission allowances attributable to U.S.-made goods.

Former U.S. Senator Larry Craig offered a forest management amendment to the Lieberman-Warner bill because he stated that the bill "would give money to Brazil to save the rain forest," but would not do enough for domestic forest conservation, where 10 million acres of forest in the United States burn each year, releasing CO_2.[18] Moreover, Senator Bernie Sanders would have banned the use of international offsets altogether in any U.S. system. Bipartisan-supported amendments in the U.S. Senate sought to eliminate the import of offsets created overseas.[19] A witness warned that leakage could occur where a company could agree to create a credit for not logging in one country, and "then just move logging operations to a neighboring country."

Forest roles continued to be a major issue in the 2010 U.S. carbon legislation. In the Waxman-Markey carbon legislation at the national level in 2009, rural interests received a significant share of legislative concessions as part of last-minute amendments to the bill that swelled its original 600 page length to more than 1,400 pages containing a number of exceptions, qualifications, and special-interest programs. Agricultural interests were successful in winning concessions transferring the responsibility for offsets from the Environmental Protection Agency, not favored by farm state representatives, to the U.S. Department of Agriculture (USDA), which has a large staff in the farm states devoted to promoting agricultural programs and is thought to favor rural offsets.

It was thought that most offsets in the United States would be in the agricultural and forest sectors. USDA has more manpower in the field and in farm states implementing different conservation and commodity programs than EPA. With the Waxman-Markey legislation in the United States, given the large share on a percentage basis of offsets that can be used by covered entities and unlimited banking of allowance provisions, emissions from regulated industries may not have to decrease under the bill until about 2030. Covered entities can use offsets for compliance and bank allowances for future use. They can then draw down the allowance bank until 2030. A recent analysis found that the bill may not actually reduce pollution at all until 2027.

The list of offset activities also was broadened to include farm interests with a provision that postpones any accounting of carbon emissions from biofuels, including corn-based ethanol. The addition of a mandatory tariff on imports from countries that do not adopt comparable controls on carbon emissions takes effect in 2020, though it may violate applicable trade treaties.

In turn, as a result of late amendments to the legislation, forest and agriculture sectors of the economy are exempted from regulations and caps. Therefore, they get a free ride. Existing biodiesel plants are exempt from lifecycle analysis under the program. It gives credit to the industries that grow biomass. It also would count as biomass residual waste materials in the paper and pulping industries.

Representative Joe Barton, a Texas Republican, referred to the deal cutting to special interests as "unprecedented" in legislation. Representative Henry Waxman, the bill's sponsor, said, "Tackling hard issues that have been ignored for years is never easy." President Obama noted about this bill, "the right balance between providing new incentives to businesses, but not giving away the store, is always an art." The jury remains out on the lasting long-term quality of this art.

Despite accounting issues and political controversy, forest preservation is the important element for inclusion in post-Kyoto carbon architecture that seeks to include developing countries. It is the means to expand the inclusion of more covered countries. The 2010 Waxman-Markey legislation in the United States includes both afforestation and prevention of deforestation as components. Debates about the role of *additionality* still prevail.

The Role of Forest Additionality

There are several issues with afforestation that revolve around so-called additionality:

- The efficiency of afforestation projects
- Monitoring and verification of reductions
- Permanence and longevity of forestation projects

Biologically based sequestration projects create issues with establishment of the appropriate business-as-usual baseline in a dynamic biologic system, the permanence of the reduction given forest change over time, and leakage of forest destruction to other terrestrial footprints elsewhere. Forestry offsets provide a challenge for preventing leakage, as logging can move to an unregulated region at another footprint, causing the net world carbon reduction impact to be zero. Quality control also can emerge as an issue with forests: legal mechanisms to ensure the value of biological credits against underperformance of credited value include insurance and bond products, buffer or surplus contingency accounts, and covenants on land-use and long-term leases and easements on use of covered land.

There is controversy as to whether such new forestation is *additional* to business-as-usual and how it is monitored and verified (see chapter 15). Some members of the environmental community contest use of forest offsets because of verification challenges. Derik Broekhoff of the World Resources Institute testified to Congress that it is extremely difficult to verify and could "undermine the integrity of the emissions cap" by qualifying offsets that would not have been additional to what would have been done anyway.[20] He also voiced concern that "leakage" could occur where a company might agree to take credit for not logging in one country, and then just move logging operations to a neighboring country or area.

Under the RGGI Model Rule in the United States, even new afforestation projects, unless insurance against biomass loss is purchased for the new forest, receive credits equal only to 90% of their calculated absorption of CO_2, to account for possible loss of forest mass over time due to fire, pest, or other causes.[21] In addition, to ensure permanent forest use under RGGI, a restrictive conservation easement is required for new forest projects that create credits. Addressing longevity in addition to amount, under RGGI, most offset credits have a 10-year period, eligible for a second 10-year period. However, new afforestation projects are eligible for a 60-year credit period. This substantially favors the net present value of afforestation projects.

By comparison, some Kyoto CERs related to forestry projects are deemed temporary for a period up to 60 years, subject to verification on a recurring 5-year basis that burning or logging do not later release carbon from the forest.[22] In the international Kyoto Protocol, forestation eligibility treatment is similar to that of RGGI and the European Union: preservation of existing forests does not qualify; adding new forests does. Kyoto's CDM only counts projects that involve planting forest in areas that were deforested before 1990, as well as afforestation, defined as planting forest in areas where there was previously no forest vegetation for at least 50 years. California is exploring reforestation projects in

Mexico to comply with California's imminent carbon GHG cap-and-trade program. Verification across borders will not be without controversy, notwithstanding the North American Free Trade Agreement (NAFTA).

So forest preservation, the natural biological mechanism for carbon absorption, has become the legal orphan of all U.S., EU, and Kyoto carbon programs. Here, additionality becomes a double-edged sword that ignores preventing deforestation, which is proceeding at alarming rates. Eligible actual reforestation does not compensate even in the order of magnitude of the much greater forest loss. Additionality applied to new afforestation only is not preserving the best natural renewable resources or a substantial enough quantity to hold the biologic resource constant.

Notes

1 Gore, Al. 2007. Speech accepting Nobel Peace Prize, Oslo, Norway, Dec. 10.

2 U.S. Environmental Protection Agency. 2008. *Draft Inventory Of U.S. Greenhouse Gas Emissions and Sinks: 1990-2006*. Federal Register. 73 (46).

3 Ingerson, Ann and Wendy Loya. 2008. *Measuring Forest Carbon: Strength and Weaknesses of Available Tools*. Wilderness Society.

4 See, BBC News. 2008. Brazil Amazon Deforestation Soars. *BBC News*. January 24. http://news.bbc.co.uk/2/hi/americas/7206165.stm; see also Remer, Lorraine. NASA Tropical Deforestation Research, Earth Observatory. http://earthobservatory.nasa.gov/Library/Deforestation/deforestation_update4.html. This source describes methods of reliably assessing deforestation.

5 Max, Arthur. 2008. Deal Struck on Forest in Climate Talks. *Associated Press*. December 9.

6 See, Carbon People. 2007. *The Problem*. http://www.carbonpeople.com.au/global_problem.html.

7 Mitchell, A. et al. 2007. *Forests First in the Fight Against Global Climate Change*. Global Canopy Programme. (June): 9. http://www.globalcanopy.org/main.php?m=121.

8 Diamond, Jared. 2005. *Collapse*. New York: Penguin Books, 471.

9 Lovejoy, Thomas. 2008. The Threat From Trees, Global Warming Isn't Just a Problem of Cars and Smokestacks But of the Chain Saw, Too. *Newsweek*. (July 7-14). http://www.newsweek.com/id/143691.

10 See Krismantari, Ika. 2008. Govt to Tender Out Financing of Five Coal-fired Power Plants. *Jakarta Post*. March 19. http://www.thejakartapost.com/news/2008/03/19/govt-tender-out-financing-five-coalfired-power-plants.html.

11 Caldeira, Ken. 2007. When Being Green Raises the Heat. *New York Times*. January 16. http://www.nytimes.com/2007/01/16/opinion/16caldeira.html?_r=1&oref=slogin;

See also, Specter, Michael. 2008. BIG FOOT: In Measuring Carbon Emissions, It's Easy to Confuse Morality and Science. *The New Yorker*. February 25. http://www.newyorker.com/reporting/2008/02/25/080225fa_fact_specter.

12 Hansen, James et al. 2008. *Target Atmospheric CO_2: Where Should Humanity Aim?* Working paper.

13 Kyoto Protocol to the United Nations Framework Convention on Climate Change. 1998. Art. 3, sec. 37 I.L.M. 22.

14 Richards, Kenneth et al. 2006. *Agriculture and Forestlands: U.S. Carbon Policy Strategies*. Pew Center on Global Climate Change. (September): 13. http://www.pewclimate.org/global-warming-in-depth/all_reports/ag_forestlands.

15 Working Group II Contribution to the Intergovernmental Panel on Climate Change. 2007. Climate Change 2007: Climate Change Impacts, Adaptation and Vulnerability. *IPCC Fourth Assessment Report*. 543; U.S. Congress. Congressional Research Service. 2008. *The Role of Offsets in a Greenhouse Gas Emissions Cap-and-Trade Program: Potential Benefits and Concerns*. RL34436, April 4: CRS 11-12.

16 Science Daily. 2008. Old Growth Forests Are Valuable Carbon Sinks. *Science Daily*. September 14. www.sciencedaily.com.

17 See, Perry, Michael. 2008. Untouched Forests Store 3 Times More Carbon – Study. *Reuters*. August 4. http://www.reuters.com/article/latestCrisis/idUSSP255954.

18 Carbon Control News. 2008. Climate Bill Amendments Roundup. *carboncontrolnews.com*. June 4.

19 Carbon Control News. 2008. Senate Climate Debate's Focus on Costs Poses Political Risks for Both Sides. *carboncontrolnews.com*. June 2.

20 Carbon Control News. 2008. Senate Panel Urged to Expand Agricultural Offsets in Lieberman-Warner Bill. *carboncontrolnews.com*. May 27.

21 Regional Greenhouse Gas Initiative. 2005. *Memorandum of Understanding*. Sec. 10.5(c)(4)(iii).

22 See, UNFCCC. 2006. Decision 5/CMP.1, UN Doc FCCC/KP/CMP/2005/8/Add.1; Watchman, Paul. 2008. *Climate Change: A Guide to Carbon Law and Practice*. London: Globe Law and Business. 96, n. 7.

21 THE FINAL ANALYSIS: THE CONCLUSION ON CARBON

According to Rajendra Pachauri, chairman of the IPCC, the lead United Nations international scientific body addressing global warming, "What we do in the next two to three years will determine our future." During the next few years, developing nations are choosing whether to deploy conventional fossil-fuel or sustainable renewable options to generate electricity. Once installed, those facilities will remain in place for 40 years or longer. These choices in energy technology made now certainly will be the signature of our carbon footprint during the crucial period of the next half century during which we may pass the point of no return in terms of global warming, as set forth in chapter 4.

Some of this imminent power production expenditure pattern reflects the very structure of the Kyoto cap-and-trade system. The Kyoto Protocol does not fundamentally change capital expenditures on power generation in the 38 Annex I nations that are covered by carbon caps, or certainly in the other 160 nations of the world that are not covered by the protocol's cap. There is no specific provision for the power generation sector. Kyoto does not specifically address the need to shift to greater reliance on non-fossil fuel sources of power generation.[1] Rather, it attempts to set national carbon allowances for larger monitored carbon sources in the 38 voluntarily participating developed countries and allows additional credits to be generated for, or sold to these sources, by offsetting carbon through various additional voluntary CDM projects in developing nations.

The Pending U.S. National Carbon System

Between now and 2012, the Waxman-Markey legislation can become the vehicle for addressing carbon at the federal level during the Obama administration. The Waxman-Markey legislation regulates carbon emissions through a cap-and-trade system, creates a national requirement for the increased use of renewable energy, and promotes a smart grid for transmission of electric power, among the primary provisions.

The Waxman-Markey carbon program would reduce annual emissions by 3% by 2012 against a 2005 baseline, 17% by 2020, and 83% by 2050. It would do so by regulating emitters of 25,000 tons CO_2e in selected industries, while exempting forestry, agriculture, and other industries from the cap. It would do so by the auction or free allocation of tradable emission allowances, with specific expenditure goals for the billions of dollars annually expected to be garnered from these auctions. The bill was changed from the initial auction of most allowances, to the free allocation of most allowances, with the number of free allowances being phased out in favor of greater auction of allowances between 2027 and 2030. Initially, 85% of allowances will be allocated for free distribution to utilities, merchant power generation facilities, steel, iron, paper, cement, refinery, and other competition-sensitive businesses. By 2030, 70% of allowances would be auctioned. The tools to determine the free allocation of hundreds of billions of dollars of free allowances over the next 20 years are controversial.

The formula for allocation of these hundreds of billions of dollars of allowances has become extremely controversial. Questions remain as to whether the formulae for distribution within an industry will be based on industry averages, specific historic emissions of a particular company, volume of sales, or a hybrid combination of the above tools. These allowances could be traded, with some specific companies forecast to receive as many as $1 billion annually of surplus free allowances above and beyond what they require. So-called early reduction credits can be earned for carbon reductions made between 2001 and 2009. State carbon control programs in 23 U.S. states could be preempted for a temporary five-year period between 2012 and 2017. Compliance could be met with offsets up to almost one-third of the number of required allowances. There remains significant controversy about the efficacy of offsets, their verification, and where and how these tools can be created and traded.

The Waxman-Markey legislation would require 6% of retail electricity to come from renewable power sources by 2012, rising to 20% by 2020.

This U.S. system will include an additional requirement to deploy energy efficiency measures to meet 1% of requirements by 2012, and 15% by 2020. The almost 30 existing state programs requiring renewable energy in the retail electric portfolios of power sellers will not be preempted by upcoming federal programs.

Legal and Policy Issues Emerging from the Carbon Toolbox

The Kyoto Protocol needs some reworking to become effective. A targeted plan of fundamental changes was outlined in part III of this book. Similar issues are appearing in the United States as the country begins to regulate climate change emissions and variables. The policy and regulatory issues quickly also become legal matters.

First, because states do not want the carbon reduction costs they impose on their in-state generators to attract higher-carbon power from out-of-state power imports, they seek to secure the borders, or at least surcharge and dissuade the intruding power flows. Because the states are attempting to not only regulate carbon produced within their borders, but also create carbon-regulated islands into which externally produced high-carbon wholesale power can no longer enter freely without penalty, there are significant Commerce Clause issues to consider. While it is perfectly understandable why certain states see this as a policy imperative, their actions trip over historic legal prohibitions against impeding the free flow of interstate commerce based on the geographic point of origin of that commerce. Wholesale electricity moves in interstate commerce at the speed of light, notwithstanding supposed "contract paths."

Second, the decision of all of the RGGI states to maximize associated revenues by auctioning all of their newly created allocations to emit carbon triggers Supremacy Clause concerns in the United States. Again here, the motives may be worthy: public money is limited, carbon emissions loom large on the policy landscape, and auctioning allocations to emit carbon maximizes public income and resources while rationing the emissions. The motive appears even more integrated when states propose to utilize the revenues of this allocation sale to fund a variety of programs that will reduce greenhouse gas production within the state.

However, jurisprudentially, motive matters according to the U.S. Supreme Court. States have expressed their purpose of this auction, to increase the price for certain high-emitting carbon power plant operations (coal in particular), as a way to change the dispatch order of which plants are allowed to run by the FERC-regulated regional independent system operators in New England, New York, California, and elsewhere. The announced objective is to make the operation of certain high-carbon-emitting plants so expensive that they become the last plants called on to operate by the regional ISO.

When unit dispatch order and operation, solely a function of federally jurisdictional terms and pricing in modern U.S. electricity markets, is manipulated indirectly by states that attempt to inflate the federally approved wholesale price at which certain facilities operate, it becomes constitutionally suspect under the Supremacy Clause of the U.S. Constitution. There are ignored, more direct, and less legally burdensome ways to get at these carbon issues, such as by requiring a certain percentage of renewable energy or low-carbon energy in the power supply mix. So the probability of finding a state legally overreaching this exclusive wholesale jurisdiction of the FERC to regulate wholesale power pricing is more likely. These particular options for renewable power promotion, either through policy incentives or natural (forest) biologic mechanisms, were surveyed in part IV.

Nowhere is the line of demarcation of federal-state responsibility pursuant to the Supremacy Clause more firmly etched in the legal precedent than in power sector regulation in the United States. In fact, Supremacy Clause jurisprudence in the power sector has its own distinct nomenclature—the filed rate doctrine. This bright line between federal and state jurisdiction has been firmly and consistently carved in the judicial firmament over three-quarters of a century of U.S. Supreme Court decisions, without variance.

Finally, but perhaps less clear, are constitutional issues with the Compact Clause and foreign commerce powers. These have already been raised in early litigation contesting the New York RGGI system in the United States. These undercut legal confidence in, and regulatory certainty of, state carbon regulation programs for global climate. California's joining with other western states and some Canadian provinces raises issues of state role in foreign commerce issues. This is not to mention a host of other legal issues that some of the state carbon regulatory initiatives may run afoul of in state statutes.

Having highlighted legal issues accompanying the particular way the leading states have embarked on carbon regulation, nonetheless, intelligent carbon policy is imperative. The consensus of scientific opinion is that this is the preeminent ecological issue of this century.

Some of the most respected climatologists argue that we have until 2015 to radically reduce the emission of CO_2, or face a very different planet, as set forth in chapter 4. We require an expedited, targeted carbon policy to temper the "goblet of fire" that powers industrial society.

But it does little to accelerate carbon restrictions at the state level, only to walk into protracted litigation that will reduce or halt the implementation of these initiatives. Whether the constitutional issues raised by the structure chosen for these state carbon schemes in the United States will prove to be fatal, is only one consideration. The corollary consideration is that the constitutional issues are real enough not to be easily dismissed by a court, and thus guarantee years of litigation and appeal, during which time these programs will be stunted, if not enjoined altogether. And if James Hansen is correct that carbon emissions worldwide must be reversed radically by 2015, then we will expend the majority of these precious remaining years litigating the legality of state carbon schemes in the United States instead of implementing legally sound solutions.

The United States is a critical player in climate control. It gets all of the recent attention, and its participation is critically important. Notwithstanding this fact, part I discussed the equally critical, and accelerating, importance of developing nations currently dealt out of all carbon control regulation. The developed world cannot control global warming effectively without the active participation of that 80% of world nations that are still developing, even though they are dealt out of the Kyoto Protocol. Without the reform of Kyoto, as set forth in part III, there is no win-win outcome for world climate control.

The Policy Response

These Commerce Clause and Supremacy Clause constitutional issues under U.S. law are just now beginning to be raised by stakeholders. Either as a corollary, or alternative strategy, to regulation of GHGs, the positive side strategies of renewable portfolio standard programs that half the U.S. states have adopted as matters of state law, and feed-in tariffs in foreign countries, are critical policy considerations. The existing U.S. state renewable programs have escaped legal challenge until mid-2008 for reasons that the carbon regulation programs will not:

- The state trust fund programs are financed by taxes in the retail utility bill to all ratepayers, and are spread across the utility consumer rate-base so that the impact on any stakeholder is de minimus at a few cents each month.

- Carbon regulation will impose huge financial obligations differentially on different generating sources, depending on their carbon emissions and size. While the impact is vested only on a limited number of stakeholders, that impact is significant and dramatically shifts the playing field for electric power production, fuel source, and economic viability in deregulated markets.
- Carbon regulation in the RGGI region will only target CO_2 emissions from larger power generating sources; it exempts smaller units and exempts other GHGs that are up to several thousand times more potent per molecule in causing global warming and persist in the atmosphere much longer than CO_2.
- There will be significant resistance to the "stick" of carbon taxes and allowance auctions imposed on previously unregulated independent power producers.
- It is unprecedented that government in the United States charges targeted regulated entities for allowances to emit air emissions.

The stakes and stakeholders are very different in carbon regulation and renewable energy programs. Both Governor Schwarzenegger's energy advisor and industry groups looking at RGGI implementation forecast litigation. In sum, the state scheme for carbon regulation, once the program regulations become final, is sure to be challenged legally by the adversely affected stakeholders, as set forth in part III.

In the end, it may be that federal carbon legislation is necessary not only for uniformity and certainty, but to eliminate the issues of the Commerce Clause, Supremacy Clause, and Compact Clause in state-formulated carbon regulation. If RGGI really was an effort to get the federal government to take more definitive action to regulate carbon emissions, then it has been effective. The 2009 proposed Waxman-Markey energy package would preempt and nullify all state carbon regulation between 2012 and 2017. Therefore, it would allow RGGI to operate for three years between 2009 and 2011, but then cause it to sunset for at least six years. California's program, which would begin in 2012, and the midwestern and western state regional carbon programs, would be preempted before they begin. Therefore, RGGI would be the one model of carbon regulation in the United States, prior to federal preemptive carbon action, if this legislation is enacted. Its impact would be significant, and perhaps controversial, as the test run of the first auction of emission allowances in the world, which may inspire the European Union countries to follow this new allowance auction technique after 2012. In fact, U.S. carbon regulation already has had international repercussions.

There is no stated rationale as to why this particular six-year period creates a hole in the doughnut of state or regional carbon regulation. However, with RGGI encountering constitutional challenges, as it already is in the first successful litigation brought against New York's implementation of RGGI, being the test model for carbon regulation in the United States, and prompting nationwide U.S. carbon regulation may be its lasting achievement. All of these issues arise only because carbon regulation is being implemented differentially by state action, which in some instances may overreach the limits of state authority or set up geographically based fences. And here may be the compelling reason for prompt enactment of carbon legislation at the federal level—it will result in immediate national policy rather than years of litigation.

Buried in a late amendment to the Waxman-Markey carbon legislation is a provision that, in a paragraph, upends the Federal Power Act demarcation between state retail rate authority and federal wholesale power jurisdiction. It excerpts, from this 75-year constant of regulatory authority over power markets, any state efforts to provide incentives to pay more for preferred sources of energy than is set by competitive power markets. Quite probably, few of the 435 members of the House read carefully all 700 pages of amendments and giveaways, as these changes have been called by numerous commentators, nor could they testify as to the purpose of this amendment. Certainly, the purpose of this provision was to remove the possible illegality of already beginning state incentives for a range of wholesale power resources that could violate the filed rate doctrine.

Federal carbon regulation avoids the challenge to individual state actions and constitutional issues. The key is formulating and implementing a workable and very prompt policy response for all nations. Richard Bradley, division head at the International Energy Agency in Paris, at the 2007 Bali Conference on the Kyoto Protocol concluded the following:

> Fossil fuel will dominate the energy supply for the foreseeable future. Investors need an international cost-effective framework if energy climate change objectives are to be met. It is not possible to discern a human influence on the emissions pathway. Policy effort is insufficient.[2]

Public finance and a workable renewable program template are necessary to build human and institutional capacity for a lower-carbon, renewable-power base in developing countries for carbon mitigation and adaptation. Fortunately, there is such a template of how to accomplish a shift to renewable power resources in developing countries, as set forth in chapter 19.

The Impact on Electric Power

In a speech I gave five years ago at Duke University, I identified 12 trends that would change the future of electric power production and use in the 21st century in the United States. They were the following:

- Increasing vulnerability to the supply of fossil fuels, including natural gas
- Depletion of supplies of economically recoverable fossil fuels
- Relative inefficiency of U.S. energy use on a global scale
- Mounting concern about environmental degradation
- Increasing concern about terrorist threats to energy security
- Vulnerability of the centralized transmission and distribution system
- Choices about whether we transport natural gas fuel or produced electricity
- The need for greater reliability of grid the system
- Differentiation of the needs for higher digital quality electricity for some uses
- Inconsistent state-level incentives for renewable power
- Deregulation and restructuring in 18 of the 50 states
- Globalization of energy markets and environmental impacts, including those of global climate change

Table 21-1 summarizes these trends. Of note on the next-to-last point, deregulation and restructuring of electric power in the United States has been frozen at the retail level since 2001 as a result of the debacle in California's electric deregulation.[3] In 2000–2001, the California power market imploded, resulting in billions of dollars of additional public debt, the bankruptcy of major utilities, and the resultant halt of all further retail deregulation across the country.

Table 21-1. Issues for the future, pivot points, and their societal forces, as of 2005[4]

Issue	Pivot Point	Type of Societal Force
1. Natural gas dependence	Increased international vulnerability	Interdependence
2. Fossil fuel depletion	Renewable energy deployment	Democratization
3. Inefficiency of U.S. energy use	Cogeneration	Decentralization
4. Environmental degradation	Renewable energy; cogeneration	Decentralization; Democratization
5. Terrorist threat	Dispersed generation and supply	Decentralization
6. T&D vulnerability	Dispersed generation	Decentralization
7. Move gas or electricity?	Dispersed generation	Decentralization
8. Need greater systems reliability	Dispersed generation	Decentralization
9. Digital electric quality	Dispersed generation or system redundancy	Mixed
10. Inconsistent state-level incentives/disincentives	New legal authority	Mixed
11. Deregulation and restructuring in only 18 states	New legal authority required	More competition
12. Global energy markets and environmental impacts	Deploy renewable technologies internationally	Democratization and decentralization

The impact of these dozen electric sector forces on society is the creation of energy *pivot points* for policymakers to respond to these forces. A pivot point is a policy or technological opportunity to turn electric production or use patterns to pivot around the issue or barrier. Many of the pivot points create more vulnerability for the energy system, decentralization of supply resources, and indicate more opportunity for renewable energy, dispersed power, and cogeneration supply. The United States is now implementing policy, both at the state and federal levels, to transition to more dispersed sources of (often renewable) power supply and more intermittent (often renewable) resources, which will have decentralizing forces on society. There is a significant push for a sustainable energy future with renewable energy and energy efficiency options, both at the state, and now federal, levels. While these technologies form the positive regulatory thrust, regulation of climate change forms the prohibitive side of the equation, regulating or penalizing certain types of power generation. The legal, policy, and regulatory issues are significant.

The basic toolbox for addressing these changes is contained in this book. I hope that you find tools that are useful as the worldwide community chisels the shape of carbon policy and renewable energy policy during the 21st century.

Notes

1 See Kyoto Protocol to the United Nations Framework Convention on Climate Change. 1998. Art. 2 and 10.

2 Bradley, Richard A. *Time is Not on Our Side: Climate Change Mitigation as an Investment Challenge.* International Energy Agency slide presentation. http://www.iea.org/speech/2007/Bradley_IPCC_Bali_Side-Event2.pdf. 16.

3 See, Ferrey, Steven. 2004. Soft Paths, Hard Choices: Environmental Lessons in the Aftermath of California's Electric Deregulation Debacle. *Virginia Environmental Law Journal.* 23 (2): 251.

4 Table adapted from Ferrey, Steven. 2005. Power Future. *Duke Environmental Law & Policy Review.* 15 (2): 261.

APPENDIX: ABBREVIATIONS

AAU	assigned amount unit
AB	Assembly Bill
APERC	Andhra Pradesh Electricity Regulatory Commission
BC	black carbon
CAA	Clean Air Act
CAFO	concentrated animal feeding operations
CAIR	Clean Air Interstate Rule
CARB	California Air Resources Board
CCX	Chicago Climate Exchange
CDM	Clean Development Mechanism
CEB	Ceylon Electricity Board
CEQA	California Environmental Quality Act
CER	certified emission reduction
CFC	chlorofluorocarbon
CFI	Carbon Financial Instrument
CH_4	methane
CO_2	carbon dioxide
CO_2e	carbon dioxide equivalent
CPUC	California Public Utilities Commission
DNA	Designated National Authority

DPUC	Department of Public Utility Control
DSM	demand-side management
EGAT	Electricity Generation Authority of Thailand
EIA	Energy Information Administration
EPA	Environmental Protection Agency
ERC	emission reduction credit
ERU	emission reduction unit
EU-ETS	European Union Emission Trading Scheme
EUA	European Union allowance
Ex-Im	Export-Import Bank of the United States
FERC	Federal Energy Regulatory Commission
GAO	Government Accountability Office
GHG	greenhouse gas
GIS	Generation Information System
GNP	gross national product
Gt	gigatons
GW	gigawatts
GWP	global warming potential
HCFC	chlorodifluoromethane
HFC	hydrofluorocarbon
IPCC	Intergovernmental Panel on Climate Change
IPP	independent power producer
ISO	independent system operator
JI	joint implementation
KV	kilovolts
kWh	kilowatt-hour
LADWP	Los Angeles Department of Water and Power
LSE	load-serving entity
LULUCF	land use and land-use change and forestry
MAC	Market Advisory Committee
MNES	Ministry of Non-Conventional Energy Sources
MOU	memorandum of understanding
MSW	municipal solid waste
Mt	million tons

MTBE	methyl tertiary butyl ether
MW	megawatts
MWh	megawatt hour
N_2O	nitrous oxide
NEG	net energy generation
NEPA	National Environmental Policy Act
NEPOOL	New England Power Pool
NERC	North American Electric Reliability Council
NGO	nongovernmental organization
NO_x	nitrogen oxide
NYDEC	New York Department of Environmental Conservation
NYISO	New York State Independent System Operator
OECD	Organization for Economic Cooperation and Development
OPIC	Overseas Private Investment Corporation
PCF	Prototype Carbon Fund
PFC	perfluorocarbon
ppm	parts per million
PSEG	Public Service Electric & Gas Company
PUC	Public utilities commission
PURPA	Public Utility Regulatory Policies Act
PV	photovoltaic
QF	qualifying facility
REC	renewable energy credit
RECLAIM	Regional Clean Air Incentives Market
RGGI	Regional Greenhouse Gas Initiative
RMU	registry removal unit
RPS	renewable portfolio standards
RTO	regional transmission organization
RUS	Rural Utilities Service
SF_6	sulfur hexafluoride
SO_2	sulfur dioxide
SPP	small power producer
SWG	Staff Working Group
TWh	terawatt hour

UN	United Nations
UNEP	United Nations Environment Programme
USDA	U.S. Department of Agriculture
WCI	Western Climate Initiative
WTO	World Trade Organization
WWF	World Wildlife Fund

INDEX

A

acid rain program, 205
adaptation policy, 143
additionality. *See also* legal additionality requirements for carbon offsets
    described, 203–204
    electric power sector and, 29
    estimation of effects of, 207
    and forestry related projects, 284–286
    history, 205
    monitoring and verification of, 206–208
    in offsets, 217
    problems with, 10
    qualification, 209–210
    of renewable power, 55, 207
    test of, 203, 204
additionality qualification, 209, 210
additionality requirements, 66, 87
additionality role of forests, 284–286
adiptic acid, 15
advanced energy facilities tax credit, 147
aerosols, 18
AES Company, 140
affirmative federal monitoring obligation over state actions, 196–197
afforestation, 281, 282, 284
afforestation projects, 85–86
Africa, 279
agricultural methane, 104
airports, 97
albedo, 18

allocations
    of allowances, 7, 220–221
    vs. auctions, 11, 68, 187–190, 290
    of emission rights, 68
allowances, 93. *See also* allocations; auctions; carbon allowances
    auction design, 170
    auction mechanisms, 295
    availability of, 169
    banking of, 63
    costs of, 169, 191–192
    early, 209
    vs. offsets, 87, 221
    prices of, 64, 86
    programs for, 63
    RGGI (Regional Greenhouse Gas Initiative), 207, 209
    taxation of trades, 64
    trading costs, 192
American Electric Power (AEP) Interstate Project, 159, 170
Andhra Pradesh, India, 269–271
Andhra Pradesh state electricity board APSEB, 270
Annex B countries, 54
Annex I countries, 51, 54, 55, 56, 61, 62, 70, 210
    power generation costs, 289
anthropogenic CO_2 emissions, 29, 35, 44
anthropogenic emission of greenhouse gas (GHG), 44
Arizona, 260
Arizona Corporation Commission, 160
Artic melting, 19–20
Asian development, 35–37

Asian population, 44
Assembly Bill 32 (AB32), 91–92, 95, 97, 99
assigned amount unit (AAU), 56, 62
atmospheric CO2 (carbon dioxide) levels, 47
atmospheric greenhouse gas (GHG) concentration, 13
auction power pricing mechanics, 187–189
auction results, 192–193
auctions. *See also* under carbon allowance auction
    vs. allocation, 11, 68, 187–190, 290
    of allowances, 11, 82, 84, 96–97, 181–182
    carbon control, 68–75
    carbon leakage and the Commerce Clause, 291
    in United States, 65
Australia, 62, 70
Austria, 73
avoided costs, 239, 240–241

B

backup/peaking power, 151, 153, 155, 156
banking of allowances, 63, 81–82
Bartlett, Brent, 153
Barton, Joe, 222, 284
baseload generation, 97, 151
Belgium, 247
Berlusconi, Silvio, 75, 128
biodiesel, 284
biofuels, 222, 284
biogas resources, 249
biological carbon sponges, 277–279
biomass projects, 85–86, 249, 250, 256, 265, 267, 268–269, 271–272, 284
black carbon (BC), 18
blackouts and brownouts, 150, 152
Blair, Tony, 37
Bonneville Power Administration (BPA), 147
Bonn International Conference on Renewable Energies, 37
Bradley, Richard, 295
Brazil, 229, 278, 279
bright line, 70
British Columbia, 106
Broekhoff, Derik, 285
Brown, Jerry, 241
Budge, Richard, 72
Bulgaria, 75, 128

C

Cabraal, Anil, 142, 264
California, 11, 64, 72, 83, 106, 107
    airports, 97
    cap-and-trade proposal, 71
    cap-and-trade system, 93
    carbon control regulation, 88
    carbon emission allowances, 69
    carbon emission standards, 97–100
    carbon regulation approach, 91–97, 99
    coal generation, 97
    coal generation imports, 73
    decoupling, 161
    electric market, 94
    feed-in tariff, 241
    feed-in tariff for renewable energy, 232
    geographic program restrictions, 257
    Independent Energy Producers Association v. California Public Utilities Commission, 240
    international offsets, 219
    locationally constrained areas, 158
    LSE (load-serving entity), 97
    noncompliance penalties, 252
    out-of-state electricity imports, 107
    out-of state generation for RPS, 257
    out-of-state offset credits, 210
    peak demand, 91
    power sector, 74
    REC banking, 251
    REC ownership, 259
    reforestation projects in Mexico, 285–286
    renewable energy projects, 260
    renewable power pricing, 196
    retail deregulation, 196, 296
    RPS target, 249
    seaports, 97–98
    WCI legal construct, 107
    Western Area Power Administration exports to, 170
    wholesale power markets, 179
California Air Resources Board (CARB), 92, 94, 96, 97, 189
California Energy Commission, 97
California Environmental Quality Act (CEQA), 136
California GHG regulation, 91–100
California Global Warming Solutions ACT (2006), 91
California Independent System Operator (CAL-ISO), 188, 292
California Public Utilities Commission (CPUC), 96, 97, 160, 196, 232, 240
California utilities, 107
California v. EPA, 141

INDEX

California v. General Motors Corporation, 137
Cambridge Energy Research Associates, 160
Canada, 67, 70, 106, 127
cap-and-trade system
    allowance auctions, 96–97
    for carbon emissions, 290
    for carbon reduction, 260
    described, 10
    in European Union, 66
    Kyoto Protocol, 51
    LADWP (Los Angeles Department of Water and Power) on, 96
    from opt-out option, 97
    opt-out option from, 97
    regional, 79
    RGGI (Regional Greenhouse Gas Initiative), 88, 187
    RGGI MOU, 80
    in United States, 205
    Waxman-Markey legislation, 7
carbon accumulation and conversion role of forests, 277–279
carbon achievement monitoring, 113–114
carbon allowances
    creation and auction, 82–84
    reservation of, 82
carbon allowances, auction, regulatory and legal issues, 187–198
    federal preemption of state carbon regulation, 187, 193–195
    filed rate doctrine, 195–197
    motive and program design, 187–193
    reserved state legal discretion, 197–198
carbon control regulation, 95
carbon controls legal comparison, 174–176
carbon controls legal distinctions, 175–179
    carbon subsides *vs.* other products, 177
    environmental protection rationales, 177–179
    ISO wholesale role, 179
    market segmentation and energy services, 177–178
    in versus out issue, 177–178
carbon cycle, 13
carbon dioxide (CO_2), 14
    atmospheric emission concentrations, 43
    emission rate, 21–22, 27, 35
    emissions in non-RGGI states, 168
    regulation by EPA, 133
    from vehicles, 141
carbon emission baselines, 75
carbon emission rights, 73
carbon emissions, 13, 181–182
Carbon Financial Instrument (CFI), 103, 104
carbon footprint, 27
carbon leakage, 171, 206, 285

carbon leakage and the Commerce Clause
    allowance auctions, 291
    carbon controls legal comparison, 174–176
    carbon controls legal distinctions, 175–179
    Commerce Clause constitutional requirements, 172–174
    leakage and policy choices, 167–172
    saving doctrine under Commerce Clause, 179–182
carbon offsets, 85–88
carbon programs. *See* carbon regulation
carbon reduction program, 42
carbon reductions, 114–116
carbon reduction targets, 294
carbon regulation, 103–105
    impacts of, 294
    regional and voluntary, 103–128
    schemes for, 212
    state level, 5, 72, 294
    in United States, 9, 106–109
carbon regulation approach, 91–92
    advisory committee changes, 93–94
    Assembly Bill 32 (AB32), 91–92, 95
    legal distinction, 94–97
carbon removal units (RMUs), 56
carbon scheme meta-screen, 212–213
carbon sequestration, 103, 279
carbon subsides *vs.* other products, 177
Carmichael, G., 18
CDM (Clean Development Mechanism). *See* Clean Development Mechanism (CDM)
Center for Biological Diversity, 143
Center for Biological Diversity, the Sierra Club, the Attorney General and San Bernardino Valley Audubon Society v. San Bernardino County, 136
Center for Biological Diversity v. Brennan, 142
Center for Biological Diversity v. Kempthorne, 142
Center for Biological Diversity v. Rural Utilities Service, 142
Central European states, 128
Ceres, 143
certainly impending injury standard, 136
certified emission reductions (CERs), 16, 52–53, 56–57, 63–65, 204
Ceylon Electricity Board (CEB), 265
Chicago Climate Exchange (CCX), 103, 104

China
    coal generation, 19, 35, 264
    coal reserves, 36, 222, 264
    electric power generation plans, 35
    emission predictions, 36
    emissions from, 133
    energy consumption and emissions, 34
    energy consumption growth, 222
    energy demand growth, 37, 45
    GHG (greenhouse gas) emissions, 35, 46
    as largest CO_2 emitter, 35
    renewable power committments, 38
    renewable power development, 264
chlorofluorocarbon (CFC), 15
Chu, Steven, 10
City of Philadelphia v. New Jersey, 178-179
Clean Air Act (CAA), 133, 205
Clean Air Interstate Rule (CAIR), 205
Clean Development Mechanism (CDM)
    additionality requirements, 204
    carbon reductions vs. renewable energy, 263
    emission reductions, 67
    forestry related projects under, 285
    offset credits, 63, 86, 107, 225
    project based offsets, 206
    projects dominating, 15
    subsidy from, 16-17
Clean Development Mechanism (CDM) projects, 54, 55, 224
clearing price, 189-190
Climate Registry, 107
coal, security from, 72
coal gasification, 249
coal generation, 11, 22
    California, 97
    embedded, 72-73
    emissions from, 136
    leakage from, 169
coal generation plans, 36, 279
coal mine methane, 104
cogeneration, 149-150, 249, 265
cogeneration projects, 268
Coke Oven Environmental Task Force v. EPA, 136
Colorado, 257
Comer v. Murphy Oil USA, 138
Commerce Clause. See also carbon leakage and the Commerce Clause
    allowances sales and, 189
    California imports, 96
    constitutional requirements, 172-174
    discriminatory regulation, 172
    geographic program restrictions, 257
    harboring wholesale power and, 160
    instate RECs, 256
    interstate commerce, 171
    out-of-state offset credits, 210

RPS (renewable portfolio standards), 293
    state feed-in tariff proposals, 233
commercial risk, 56
Compact Clause, 73, 100, 292, 294
compensatory tax doctrine, 179-180
Competitive Renewable Energy Zones, 158
compliance mechanisms, 127
compliance operations, 218
compliance payments, 252
conclusions
    about, 289
    electric power impact, 296-298
    legal and policy issues, 291-293
    policy response, 293-296
    United States National Carbon System, 290-291
Congressional Research Service, 218
Connecticut, 79, 189, 252, 253
Connecticut Department of Public Utilities (DPUC), 256, 259
Connecticut Supreme Court, 259
conservation
    creating credits from, 217-226
    and renewable power, 219-222
    in the United States and RGGI Schemes, 219-222
constitutional law, 173, 187
constitutional law and states' cases
    feed-in tariff for renewable energy, 240-243
    independent energy power producers, 240
    Independent Energy Producers Association v. California Public Utilities Commission, 240
    Public Utility District No. 1 of Snohomish County Washington, 242-243
    San Diego Gas & Electric, 240-242
    Southern California Edison Company, 240-242
consumer groups vs. environmental groups, 162
contracts for differences, 63
control periods, 81
cost of energy efficiency, 156
costs
    of allowances, 191-192
    avoided, 239, 240-241
    of forest preservation, 281
    of power generation, 289
    of renewable power, 225
    of uplift, 155
Craig, Larry, 283
cyber security, 148
Czechoslovakia, 74, 128
Czech Republic, 70

D

decentralized *vs.* centralized carbon control, 73–74
deforestation, 278–279, 284
deindustrialization, 68
Delagado, Juan, 65
Delaware, 79, 251, 253, 258
demand control, 148
demand response resources, 156
demand-side management (DSM), 171, 249
demand-side management (DSM) resources, 156
Denmark, 68
deregulated power sector, 74
deregulated suppliers, 62
deregulation of retail electric utilities, 73, 94, 161, 176
Designated National Authority (DNA), 53
developing countries
    Asian renewable power development, 265
    black carbon (BC) emissions, 19
    CDM (Clean Development Mechanism) projects, 211
    compliance phase-in for, 71
    development growth curve, 273
    electric demand and production in, 272
    electrification of, 46
    energy infrastructure investment timing, 29
    Equator Principles, 225
    obligation avoidance, 51
    offsets in, 223
    power generation mix, 33
    power sector model in, 265–272
    primary energy usage in, 34–35
    world population of, 44
discriminatory regulation, 172, 180
dispatch of renewable power, 151
distributed generation, 148, 159, 239, 249, 257
District of Columbia, 85, 251, 258
Dominion Resources, 159
dormant Commerce Clause.
    See Commerce Clause
dual fuel oil/gas capability, 153–154
Duke Energy, 73
Dynegy, 190

E

early allowances, 209
early court challenge on climate change, 135–140
early reduction, 81
early reduction credits, 8, 87, 290
Earth temperature, 13
East Coast Regional Gas Initiative (RGGI), 11
Eastern European EU countries, 70, 72, 128
East Texas Electric Cooperative v. Rural Utilities Service, 141
electricity, 6
electricity deregulation, 62
Electricity Generation Authority of Thailand (EGAT), 265
electricity sector emissions estimates, 106
electric power
    demand, 32
    impact, 296–298
    production trends, 296
electric power and carbon emissions
    Asian development, 35–37
    developing countries, 34–35
    fossil fuels, role of, 27–29
    renewable power, 32–34
    renewable power and coal, 37–38
    technology impact, 29–31
Electric Reliability Council of Texas (ERCOT), 155
electrification of developing countries, 46
emission penalties, 127
emission reduction credits (ERCs), 205
emission reduction units (ERUs), 55, 56, 57, 63
emission rights, 62, 68, 177
emissions. *See also* allowances; greenhouse gases (GHGs)
    caps on, 51
    reductions, 7, 66, 67
    and sequestration, 66
endangered species, 142
energy conservation projects, 219
energy consumption and emissions, 33–34
energy demand increases, 45
energy efficiency, 156, 157
Energy Independence and Security Act of 2007, 148
Energy Information Administration (EIA), 22, 149
Energy Policy Act of 1992, 103
Energy Policy Act of 2005, 157, 159
energy storage, 148, 149, 153
energy use statistics, 45
enforcement, lack of, 128

England, 36
enlarging group of regulated countries, 117-121
Entergy Louisiana, Inc. v. Louisiana Public Service Commission, 195, 196
environmental groups
    on additionality for renewable resources, 55, 207
    on allowance auctions, 69
    on allowances, 190
    climate change/GHG (greenhouse gas) litigation, 133-136, 141-142
    vs. consumer groups, 162
    Kyoto Protocol lawsuits, 125
    on offsets, 97, 107-108
    on RGGI plans, 87
Environmental Protection Agency (EPA), 221, 283
environmental protection rationales, 177-179
Equator Principles, 225
equivalent burden exception, 180
ERUs (emission reduction units). *See* emission reduction units (ERUs)
Estonia, 71
ethanol, 284
EU-ETS (European Union Emission Trading Scheme)
    additionality requirements, 204
    allowance banking, 63
    compliance penalties, 127
    vs. Kyoto Protocol, 64
    lessons from, 66
    lessons from offsets, 224
    in RGGI (Regional Greenhouse Gas Initiative), 86
    scope of, 61
    trading, 62, 65
European Commission, 65, 74
European Union, 11, 66
    carbon program, 82-83
    carbon reduction program, 42
    European Trading scheme, 83
    feed-in tariff for renewable energy, 229
    national carbon regulation, 108
    new *vs.* old membership, 70
    policy disputes in, 68
European Union allocations, 73
European Union carbon control *vs.* U.S.
    carbon control, 61-75
    allowance auctions, 68-75
    basics of, 61-63
    coal generation, embedded, 72-73
    decentralized *vs.* centralized, 73-74
    Kyoto Protocol linkage, 63-64
    new entrants and degree of development, 70-71
    pace and scope of, 74-75
    tensions from, 67-75
    trading, 62-63

European Union Court of Justice, 74
European Union Emission Trading Scheme (EU-ETS), 11
evapotranspiraton, 21
excess profits, 191
Exchange allowances, 103
Exchange Offsets, 104
Export-Import Bank of the United States (Ex-Im), 139

F

federal agencies, 4
federal carbon regulation, 294
Federal Energy Regulatory Commission (FERC), 94, 156, 161
    authority, 159
    jurisdiction of, 179, 193-195, 198
    just and reasonable rates, 191
    PURPA (Public Utility Regulatory Policies Act), 240-241
    renewable energy pricing, 240-241
federal jurisdiction, 193
Federal permitting agencies, 136
Federal Power Act, 235, 242, 295
Federal Power Act of 1935, 172, 193, 194, 196, 233
Federal Power Act of 2005, 241
federal preemption doctrine, 187, 190, 195, 196
federal preemption of state carbon regulation, 187, 193-195, 256, 290
federal preemption *vs.* state authority, 234
federal proposals, 232
federal wholesale policy *vs.* state retail authority, 295
feed-in tariffs
    California, 241
    elements of, 229-231
    federal proposals for, 232
    in foreign countries, 293
    India, 265
    jurisdiction for, 242
    vs. renewable portfolio standards (RPS), 230-231
    Sri Lanka, 265
    state adoption, 293
    state proposals for, 232
feed-in tariffs, for renewable energy, 232-233
    constitutional law and states' cases, 240-243
    federal preemption *vs.* state authority, 234
    filed rate doctrine, 234-239
    internationally, 229-231
    in United States, 229-231

INDEX

FERC (Federal Energy Regulatory Commission). *See* Federal Energy Regulatory Commission (FERC)
Ferrey, Steven, 142, 264
filed rate doctrine, 193, 198
    affirmative federal monitoring obligation over state actions, 196–197
    California renewable energy pricing, 240
    feed-in tariff for renewable energy, 234–239
    green energy limited exemption, 235
    net metering exemption, 236–237
    schemes for, 292
    unit dispatch order, 292
    U.S. Supreme Court decisions, 195–196
first-seller approach, 93
Florida, 233
forestation carbon sequestration, 279
Forest Carbon Participation Project, 281
forests/forestry, 104, 277–286
    additionality role, 284–286
    afforestation, 281, 282, 284
    afforestation projects, 85–86
    carbon accumulation and conversion role, 277–279
    carbon emissions from, 282
    deforestation, 278–279, 284
    eligibility, new *vs.* old, 282
    international carbon regulation and forest credits, 279–283
    preservation costs, 281
    reforestation, 281, 285, 286
    related projects and additionality, 284–286
    U.S. carbon legislation and, 282–284
fossil fuels, 27–29, 36
France, 69, 72, 74
free electricity, 211, 263
Friends of the Earth v. Mosbacher, 139
fuel cells, 249
fundamental changes required, 116–125

G

Gao, Li, 127
General Agreement on Tariffs and Trade (GATT), 71, 127
generation unavailability, 152
geothermal power, 249, 250
Germany, 36, 72, 231
Global Change Research Act, 142
Global GHG market, 93
global warming
    carbon-induced, 13
    effects of, 133
    impact of, 21
    renewable energy and, 143
global warming, court rulings, 133–144
    early court challenge on climate change, 135–140
    recent legal claims, 140–144
    U.S. Supreme Court Ruling, 134–135
goals and regulatory mechanisms, 79–81
Godmanis, Ivars, 71
goods *vs.* services, 177
Gore, Al, 20, 37
Government Accountability Office (GAO), 211, 224
Greece, 70, 71, 72
Green Communities Act, 161
green energy limited exemption, 235
green grid, 109
greenhouse gases (GHGs), 6, 14–17
    anthropogenic emission of, 44
    atmospheric concentration, 13
    cap-and-trade system for, 66
    carbon regulation, 197
    emissions, 67
    emissions in China, 46
    forgotten and ignored, 18–19
    levels of, 22
Greenland, 20
grid connections, 150
grid operator, 154

H

Hansen, James, 18, 37, 43, 293
Hawaii, 232, 257
HCFC-22, 16
HFC-23. *See* trifluoromethane (HFC-23)
HFCs (hydrofluorocarbons) reduction, 17
high-carbon power leakage, 95
Holdren, John, 43
human-made climate change, skepticism regarding, 71
Hungary, 68, 70, 74, 128
Hurricane Katrina Litigation, 138
hydrofluorocarbon (HFC), 14, 204
hydro resources, 249

I

Iceland, 63, 66
Idaho PUC, 259
Illinois, 108, 232, 257
imported power, 93, 97, 171
import tariff, 222
incidental burden standard, 173, 176
Indeck-Corinth, 11
Indeck Energy, 197
independent energy power producers, 240
Independent Energy Producers Association v. California Public Utilities Commission, 240
independent power producer (IPP), 160, 190
Independent System Operators (ISOs), 83, 179, 292
India
    coal generation, 19, 264
    coal reserves, 36, 222, 264
    emission predictions, 36
    emissions from, 133
    energy consumption, 36
    energy consumption and emissions, 34
    energy demand increases, 45
    feed-in tariff, 265
    GHG (greenhouse gas) emissions, 35
    power sector model in developing nations, 269–272
    renewable power development, 264, 265
    small hydro projects, 269
Indiana, 108, 233
Indonesia
    coal generation plans, 279
    coal reserves, 36, 264
    deforestation, 278
    feed-in tariff for renewable energy, 229
    GHG (greenhouse gas) emissions, 279
industrial CO_2 emitters, 138
ineligibility of renewables, 85–88
insurance companies, 138
Inter-American Commission on Human Rights, 139
interest groups, 220
Intergovernmental Panel on Climate Change (IPCC), 13, 21
intermittent electric resources, 150, 155, 160. *See also* renewable power; solar energy; wind power
international carbon equation, 272–274
international carbon regulation and forest credits, 279–283

International Energy Agency (IEA), 22, 38, 45, 46
    energy consumption and emissions, 33, 34
international enforcement mechanisms, 125–128
international offsets, 218–219, 283
interstate commerce, 171
interstate taxation schemes, 179
Inuit petition, 139, 140
in-versus-out issue, 177–178
investor-owned utilities, 94, 96
Iowa, 108, 249, 257
IPCC (Intergovernmental Panel on Climate Change), 13, 21
IPCC Fourth Assessment Report (2007), 21, 42
Iran, 72
Ireland, 68
Israel, 229
Italy, 36, 69, 70, 72, 73, 75, 128, 247

J

Jacobson, Mark, 18
Japan, 67, 70, 127
joint implementation (JI) Kyoto credits, 107
jurisdiction, 193–195. *See also* under federal preemption
just and reasonable rates, 234

K

Kansas, 108
Kansas Department of Health, 143
Kenya, 229
Kivalina v. ExxonMobil Corp., 140, 141
Klaus, Vaclav, 71
Korea, 229
Korsinsky v. EPA, 135
Kyoto Protocol. *See also* Clean Development Mechanism (CDM)
    2012 termination of, 64
    Annex I countries, 62
    architecture of, 51–57
    basic structure, 51–52
    carbon control linkage, 63–64
    carbon trade, risk and benefits, 56–57
    developing countries and, 35
    GHG (greenhouse gas) carbon regulation, 15

INDEX 311

international trade, 71
legal additionality requirements for carbon offsets, 210–212
limitations of, 291
objectives, 5
offsets, 52–56
problems with, 9, 36, 109
reform needs, 293
Kyoto Protocol Clean Development Mechanism (CDM). *See* CDM (Clean Development Mechanism)
Kyoto Protocol critique, 113–128
    carbon achievement monitoring, 113–114
    carbon reductions, 114–116
    fundamental changes required, 116–128
Kyoto Protocol II mechanisms, 63

L

LADWP (Los Angeles Department of Water and Power). *See* Los Angeles Department of Water and Power (LADWP)
landfill gas resources, 249
Land use and land-use change and forestry (LULUCF), 279
Latvia, 71
leaded gasoline, 80
leakage, 71, 189. *See also* carbon leakage
leakage and policy choices, 167–172
leakage of carbon, 108, 109
legal additionality requirements for carbon offsets, 203–218
    additionality described, 203–204
    additionality history, 205
    additionality qualification, 209–210
    and Kyoto Protocol, 210–212
    as meta-screen in carbon scheme, 212–213
    monitoring and verification of additional carbon controls, 206–208
legal and policy issues, 291–293
legal decisions, 173. *See also* U.S. Court of Appeals; U.S. Supreme Court
    California v. EPA, 141
    California v. General Motors Corporation, 137
    Center for Biological Diversity, the Sierra Club, the Attorney General and San Bernardino Valley Audubon Society v. San Bernardino County, 136
    Center for Biological Diversity v. Brennan, 142
    Center for Biological Diversity v. Kempthorne, 142
    Center for Biological Diversity v. Rural Utilities Service, 142
    City of Philadelphia v. New Jersey, 178–179
    Coke Oven Environmental Task Force v. EPA, 136
    Comer v. Murphy Oil USA, 138
    East Texas Electric Cooperative v. Rural Utilities Service, 141
    Entergy Louisiana, Inc. v. Louisiana Public Service Commission, 195, 196
    Friends of the Earth v. Mosbacher, 139
    Independent Energy Producers Association v. California Public Utilities Commission, 240
    Kivalina v. ExxonMobil Corp., 140, 141
    Korsinsky v. EPA, 135
    Maine v. Taylor, 180–181
    Massachusetts v. EPA, 133, 134, 136, 138, 139, 140
    Montana Environmental Information Center v. Johanns, 142
    New England Power Co. v. New Hampshire, 178
    New York v. U.S. EPA, 136
    Northwest Environmental Defense Center v. Owens Corning, 138
    Pacific Gas & Electric Co. v. California Energy Resources Conservation and Development Commission, 193
    Public Utility District No. 1 of Snohomish County, Washington v. FERC, 197
    State of Connecticut v. American Electric Power, 137
    Steadfast Insurance Company v. The AES Corporation, 141
    Sunflower Electric Power Corp. v. Kansas Department of Health and Environment, 143
    United Haulers Association, Inc. v. Oneida-Herkimer Solid Waste Management Authority, 181
    West Lynn Creamery, Inc. v. Healy, 174–176
legal issues, 108–109
legislation, 74
liability risk, 133
Lichtenstein, 63, 66
Lieberman-Warner bill, 3, 282–283
Lithuania, 68, 71
litigation, 137
litigation risk, 133, 139
load serving entities (LSEs), 94, 95, 97, 99, 171

local government permitting agencies, 136
long-term power contracts, 83
Los Angeles Department of Water and Power (LADWP), 69, 71, 74, 95, 96, 97
Luxembourg, 62, 69

M

Madagascar, 278
Maine, 79, 87, 160
    noncompliance penalties, 252, 254
    REC credit prices, 253
    REC ownership, 259
    REC sources, 257
Maine v. Taylor, 180–181
Manitoba, 106, 108
Market Advisory Committee (MAC), 93, 99
market-based design, 80
market-based rates, 241
market bid rules, 196
market segmentation and energy services, 177–178
Maryland
    carbon leakage, 169
    decoupling, 161
    New England Power Pool (NEPOOL) RPS credit recognition, 253
    PJM interconnection, 258
    REC banking, 251
    REC credit prices, 253
    RGGI (Regional Greenhouse Gas Initiative), 79
    RGGI litigation risk, 73
Massachusetts
    decoupling, 161
    early allowances, 209
    New York ISO REC trades, 258
    noncompliance penalties, 252
    REC banking, 251
    REC credit prices, 253
    REC sources location origin, 257
    RGGI (Regional Greenhouse Gas Initiative), 79
    RGGI litigation risk, 73
    RPS tracking system, 257
    solar photovoltaic REC program, 236
    transmission support, 160
Massachusetts v. EPA, 133, 134, 136, 138, 139, 140
Mercury Rule, 205
methane (CH_4), 14, 15, 104
methane flaring, 204, 210, 211, 263
Mexico, 285–286

Michigan, 108, 232–233
Midwestern Greenhouse Gas Reduction Accord, 108
Midwestern states, 108
Midwest regional U.S. carbon regulation, 108–109
mini-hydro projects, 267
Ministry of Non-Conventional Energy Sources (MNES), 269
Minnesota, 108
monopolized power sector, 73
Montana Environmental Information Center v. Johanns, 142
Montreal Protocol, 71, 80
Montreal Protocol on Substances that Deplete the Ozone Layer, 15
motive and program design, 187–193
motives for allowance auctions, 190–192
municipal solid waste (MSW), 249
municipal utilities, 94, 95, 97

N

NAFTA (North American Free Trade Agreement), 286
national allocation, 68
National Allocation Plans, 65, 83
National Ambient Air Quality Standards (NAAQS), 205
National Environmental Policy Act (NEPA), 139, 141, 142
National Transmission Study, 147
Native American groups, 140
natural gas exports, 72
Nazarenko, Larissa, 18
Netherlands, the, 247
net metering, 85, 236–239, 258–259
net metering exemption, 236–237
netting, 236
Nevada, 251, 260
New England, 251
New England grid, 154, 157
New England Independent System Operator (NE-ISO), 154, 188, 292
New England Power Co. v. New Hampshire, 178
New England Power Pool (NEPOOL), 253, 257
new entrants and degree of development, 70–71

New Hampshire, 79, 252
New Jersey, 79, 236
    decoupling, 161
    New England Power Pool (NEPOOL) RPS credit recognition, 253
    noncompliance penalties, 252
    REC credit prices, 253
New Mexico, 106
new *vs.* old forest eligibility, 282
New York, 11, 69, 73, 74, 79, 87
    decoupling, 161
    GHG (greenhouse gas) carbon regulation, 197
    independent system operator (ISO), 292
    New England Power Pool (NEPOOL) RPS credit recognition, 253
    PJM REC trades, 258
    REC credit prices, 253
    REC sources, 257
    renewable energy projects, 260
New York City, 154
New York Department of Environmental Conservation (NYDEC), 190
New York RGGI system, 292
New York State Independent System Operator (NYISO), 169, 188, 258
New York v. U.S. EPA, 136
New Zealand, 62, 70
Nicaragua, 229
nitrogen oxide (NO$_x$), 127
nitrogen oxide (NO$_x$) summer budget program, 205
nitrogen trifluoride (NF$_3$), 18
nitrous oxide (N$_2$O), 14, 15
noncompliance penalties, 252
nongovernmental organization (NGO), 162
non-justiciable political question, 138, 139
non-spinning reserves, 154
non-utility generators (NUGs), 179
North American Electric Reliability Council (NERC), 151, 156
North American grid, 151
North Carolina, 257
Northwest Environmental Defense Center v. Owens Corning, 138
Norway, 63, 66, 229
nuisance claims, 140
nylon-66, 15

O

Obama, Barack, 10, 148, 221, 222, 284
Obama administration, 4, 133, 143, 147, 290
ocean tidal renewable resources, 249
off-grid energy, 37
offset allowances, 81, 85, 86
offset credits, 86, 203
offsets
    additionality qualification, 210
    vs. allowances, 87, 221
    California cap-and-trade, 93
    CDM (Clean Development Mechanism), 52–54
    CDM (Clean Development Mechanism) future, 54–55
    compliance with, 8
    credibility of, 210
    functions of, 221
    JI (joint implementation) as offset credit transfer, 55–56
    RGGI (Regional Greenhouse Gas Initiative), 85
    transfer of, 210
Offsets Aggregators, 104
Offsets Program, 104
Offsets providers, 104
offsetting carbon, 217–226
Ohio, 108
Ontario, 108
opt-out option from cap-and-trade system, 97
Oregon, 106, 236
out-of-state electricity imports, 168
out-of-state offset credits, 210
Overseas Private Investment Corporation (OPIC), 139
ozone layer, 15

P

Pachauri, Rajendra, 211
Pacific Gas & Electric Co. v. California Energy Resources Conservation and Development Commission, 193
Pataki, George, 79
peak demand, 154
peaking capacity, 154
penalty assessment, 62
Pennsylvania, 71, 108, 253
Pennsylvania-Jersey-Maryland Independent System Operator (PJM), 188

perfluorocarbon (PFC), 14
Philippines, 278
photovoltaic (PV) power, 32, 267, 269
pivot points, 296
PJM Generation Attribute Tracking System (GATS), 257
PJM interconnection, 258
Poland, 67, 69, 72, 74, 75, 128
policy disputes, 68
policy response, 293-296
political question doctrine, 137
population growth, 42
Portugal, 62
power companies, 137
power generation mix, 33
power purchase agreement (PPA), 265, 273-274
power sector model in developing nations
    about, 266-267
    India, 269-272
    international carbon equation, 272-274
    Sri Lanka, 266-267
    Thailand, 267-269
power technology changes, 121-125
preferential wholesale rates, 242
price mechanism, 189-190
price risk, 56
program design, 187-193
    auction power pricing mechanics, 187-189
    auction results, 192-193
    motives for allowance auctions, 190-192
    state battles on auctions, 189-190
proprietary actions, 181-182
Prototype Carbon Fund (PCF), 224-225
proxy values, 66
public cooperative utilities, 94
public nuisance litigation, 139
Public Service Electric & Gas Company (PSEG), 171
Public Utility Commission (PUC), 94
Public Utility District No. 1 of Snohomish County, Washington v. FERC, 197
Public Utility District No. 1 of Snohomish County Washington, 242-243
Public Utility Regulatory Policies Act (PURPA), 73, 198, 235, 240, 259

Q

Qatar, 72
quadruple dipping, 236
qualifying facilities (QFs), 198, 240
quarantine exception, 180-181
Quebec, 257
quick-start backup power, 153

R

Ramanathan, V., 18, 19
recent legal claims, 140-144
RECs. *See* renewable energy credits (RECs)
redressability requirements, 133
reforestation, 281
refrigerants, 15
regional and voluntary carbon programs, 103-128
regional cap-and-trade system, 79
Regional Clean Air Act Incentives Market (RECLAIM), 80, 205
regional CO_2 emissions budget, 81
Regional Greenhouse Gas Initiative (RGGI)
    additionality requirements, 204
    allocations *vs.* auctions, 69
    allowance point trigger threshold, 64
    allowances, 207, 209
    carbon allowances. creation and auction, 82-84
    carbon offsets, 85-88
    control under, 74
    energy supply system reliability, 169
    goals and regulatory mechanisms of, 79-81
    ineligibility of renewables, 85-88
    litigation risk, 73
    offset reduction obligation, 218
    offsets additionality requirement, 207
    and renewable energy, 219
Regional Greenhouse Gas Initiative (RGGI) states
    auction decision, 84
    vs. California carbon regulation approach, 99
    carbon leakage into, 170
    Model Rule provisions, 82-83
    MOU (memorandum of understanding), 79-80
    scope of, 108
    Supremacy Clause, 291
    wholesale power markets, 179

INDEX 315

regional transmission organization (RTO), 83, 179
regional U.S. carbon regulation, 106-109
regulatory incentives, 162
regulatory jurisdiction, 70, 74
renewable capacity additions, 250
renewable energy, 8, 104. *See also* renewable power
   global warming and, 143
   and RGGI (Regional Greenhouse Gas Initiative), 219
   subsidies and allowances, 219
renewable energy credits (RECs), 107, 232, 247
   in-state multipliers, 257
   and offsets value, 254-256
   pricing, 256
   tracking, 251
renewable portfolio standards (RPS)
   capabilities, 259-260
   expansion timing, 161
   vs. feed-in tariff, 230-231
   vs. feed-in tariffs, 231, 243
   NERC concerns regarding, 151
   for renewable power, 247-260
renewable power. *See also* feed-in tariff for renewable energy
   about, 263-264
   additionality of, 55, 203
   benefits of, 263
   and coal, 37-38
   dispatch of, 151
   emission allowances to, 63
   expansion, 225-226
   feed-in tariff, 239
   forms of, 263-264
   intermittent electric resources, 149
   net metering, 85
   power sector model in developing nations, 265-272
   relative development costs of, 225
   renewable portfolio standards (RPS) for, 247-260
   retail electricity from, 290
   role of, 32-34
   shift to, 29
   transformation to, 263-269
   transmission for, 147
   transmission issues, 158
renewable power, creating credits from, 217-226
   about, 217
   Kyoto-EU carbon system, 222-224
   Kyoto Protocol CDM limitations, 224-226
   pivotal role of carbon offsets, 218-219
   in the United States and RGGI Schemes, 219-222

renewable power, renewable portfolio standards for
   design and contours, 247-248
   renewable energy credits (RECs) and offsets value, 254-256
   RPS capabilities, 259-260
   state program legal issues, 256-259
   state program variations, 251-254
   state variations and results, 249-251
Renewable Power in Developing Countries: Winning the War on Global Warming (Ferrey and Cabraal), 142, 264
renewable power locations, 256-258
renewable power pricing, 196
renewable power trust fund, 85
renewable resource zones, 160
renewables, ineligibility of, 85-88
renewable system benefit charges, 161
retail electricity, 290. *See also* deregulation of retail electric utilities
RGGI. *See* Regional Greenhouse Gas Initiative (RGGI) states
RGGI Model Rule, 85, 86, 87, 219, 285
RGGI Schemes, 219-222
Rhode Island
   feed-in tariff for renewable energy, 233
   incentive profit legislation, 162
   instate RECs, 251
   noncompliance penalties, 252, 253
   REC credit prices, 253
   RGGI (Regional Greenhouse Gas Initiative), 79
   RGGI plan, 87
ripeness, 135
Romania, 70
RPS (renewable portfolio standards). *See* renewable portfolio standards (RPS)
rural electric cooperatives, 94
rural offsets, 221
Rural Utilities Service (RUS), 141, 142
Russia, 36, 62, 67, 72

S

Sanders, Bernie, 283
San Diego Gas & Electric, 240-242
saving doctrine under Commerce Clause
   compensatory tax doctrine, 179-180
   proprietary actions, 181-182
   quarantine exception, 180-181
Scharzeneggar. Arnold, 74, 75
seaports, 97-98

self generation, 150
Sempra Energy Utilities, 69
sequestration
    biologically based projects, 284
    and emissions, 66
Sierra Club, 141, 143
Sikorski, Radek, 71
Skoda, Jan, 72
Slovakia, 70, 72
small hydro projects, 269
small power producers (SPPs), 265
smart grid, 147–163
    construction, 147–150
    decoupling and operator incentives, 161–163
    management, 150–157
    for transmission, 290
Solar Electric Power Association, 231
solar energy, 32, 150, 160, 249, 250, 251. *See also* photovoltaic (PV) power
soot, 18
South America, 279
South Dakota, 108
Southern California Edison, 97, 159, 160, 240–242
sovereign risk, 56
Spain, 73, 231
special-interest programs, 221
species extinction, 21
spinning reserves, 154, 155
Sri Lanka, 229, 266–267
standards and outcomes, 173–174
standards of judicial review, 173–174
state battles on auctions, 189–190
state carbon control programs, 290
state carbon regulation, 72, 294
state government permitting agencies, 136
state legal discretion, 197–198
state legislative proposals, 232–233
State of Connecticut v. American Electric Power, 137
state program legal issues, 256–259
state programs, 248
state regulatory actions, 79
state renewable programs, 251–254, 256–259, 293
state retail authority *vs.* wholesale rate authority, 241, 295
state role in foreign commerce issues, 292
states feed-in tariff, 293
Steadfast Insurance Company v. The AES Corporation, 141
stimulus package, 147

strict scrutiny standard, 172, 173, 178
subsidy mechanisms, 249
substantive merits, 135
sulfur dioxide (SO_2) trading, 56, 127
sulfur hexafluoride (SF6), 14
Sunflower Electric Power Corporation, 143
Sunflower Electric Power Corp. v. Kansas Department of Health and Environment, 143
Supremacy Clause, 100, 109, 193, 235, 291, 293
surcharge and subsidy system, 177
sustainable development, 53, 211
Sweden, 247
Switzerland, 229
synchrophasor technology, 159
system benefit charges, 250

T

tailpipe standards, 107
Tamil Nadu, India, 271–272
Tamminen, Terry, 74, 187
tax and subsidy schemes, 175
taxation of allowance trades, 64
technology choices, 46–47
technology impact, 29–31
technology transfer, 211
Teflon, 15
tensions from European Union carbon control *vs.* U.S. carbon control, 67–75
test
Texas, 158, 251
Thailand
    competitive solicitation, 265, 267
    Electricity Generation Authority of Thailand (EGAT), 265
    power sector model in developing nations, 267–269
    renewable power development, 265
time-based rate schedule, 157
tipping point, 38, 72
    critical point, 41–44
    forecast, 11
    issues regarding, 41–47
    linkage and response, 44–47
    population, 44–45
    technology choices, 46–47
trade sanctions, 71
trading. *See also* cap-and-trade system
    carbon control, 62–63
    of emission rights, 62

trading prices, 63
trading system, 104
Trans-Allegheny Interstate Line (TrAIL) Project, 159, 170
transmission. *See also* smart grid
   additions to, 170
   capability, 256
   congestion, 259
   cost obligations, 160
   for distributed generation, 239
   in-state interconnections, 256
   for out-of-state REC's, 256
   reliability, 153
   renewable power issues, 158
   system efficiency, 157–158
Transmission Corporation of Andhra Pradesh Limited (APTransco), 270
transmission expansion, 157–161
transmission management
   grid reliability, 155–157
   inbalance penalties, 156
   quick-start backup power, 153–155
   renewable energy, 150–153
   transmission expansion, 157–161
transmission system, 265
transportation sector emissions, 97
trifluoromethane (HFC-23), 15, 122, 223–224
Turkey, 70, 229
Tusk, Donald, 75

U

Ukraine, 62, 67
UN Framework Convention on Climate Change, 106
unit dispatch order, 292
United Haulers Association, Inc. v. Oneida-Herkimer Solid Waste Management Authority, 181
United Kingdom, 73, 247
United Nations Environment Programme (UNEP), 22, 41
United States
   allowance auctions in, 65
   cap-and-trade system, 127
   carbon legislation, 282–284
   carbon regulation, 9, 104, 295
   climate change litigation in, 135
   coal generation in, 72
   coal reserves, 36, 264
   conservation and renewable power in, 219–222
   emissions reductions, 103
   energy consumption and emissions, 28
   federal climate change legislation, 104
   feed-in tariffs, 229–233
   Kyoto limit, 128
   national carbon regulation, 108
   regulatory schemes, 187
   renewable energy credits (RECs), 247
   SO_2 allowance programs, 63
   state carbon allowance auctions, 69
   state carbon regulation, 64
   sulfur dioxide (SO_2) trading, 53
United States carbon control *vs.* European Union carbon control, 61–75
United States National Carbon System, 290–291
uplift costs, 155
U.S. Clean Air Act, 83
U.S. Congress Government Accountability Office (GAO), 66
U.S. Constitution, 100, 171, 181, 229. *See also* constitutional law
   interstate commerce, 95–96
   Supremacy Clause, 292
U.S. Court of Appeals
   District of Columbia Circuit, 134–135, 136, 205
   Fifth Circuit, 139
   Fourth Circuit, 159
   Ninth Circuit, 137, 139, 142, 194, 195–196, 197, 234, 240, 241, 242
   Second Circuit, 138
   Southern District of New York, 135–136
U.S. Department of Agriculture (USDA), 141, 221, 283
U.S. Department of Energy, 22, 45, 103, 148, 161
U.S. Department of Interior, 135
U.S. Environmental Protection Agency (EPA), 133
U.S. National Carbon System, 7–8
U.S. Supreme Court
   on auction allowances motive, 291
   Commerce Clause, 172
   compensatory tax doctrine, 180
   discriminatory regulation, 178
   EPA authority ruling, 133, 134–135
   equivalent burden exception, 180
   on federal preemption, 234
   FERC jurisdiction, 197
   filed rate doctrine, 195–196, 235
   harboring wholesale power, 160
   preferential wholesale rates, 242
   proprietary actions, 181–182
   quarantine exception, 180–181
   standards of judicial review, 173–174
utility companies, 137

V

vehicles, 141
Vermont, 79, 176, 233
vintage allowances, 209
vintage REC ownership, 258–259
vintage RECs, 258
voluntary programs, 235

W–Z

walruses, 20
Washington State, 106, 252
waste energy projects, 267
water vapor, 14, 18
Waxman, Henry, 222
Waxman-Markey legislation, 3, 4, 7, 8, 74, 147, 151, 159, 212, 220, 231, 253, 256, 283, 284, 290, 294, 295
wealth transfer, 71
Western Area Power Administration, 95, 147
Western Area Power Administration exports, 170
Western Climate Initiative (WCI), 72, 74, 106, 107
Western European EU countries, 70
Western European states, 128
Western states, 108
West Lynn Creamery, Inc. v. Healy, 174–176
West regional U.S. carbon regulation, 106–107
wholesale power definition, 194
wholesale power market, 172
wholesale rate authority *vs.* state retail authority, 241
windfall profits, 191, 259
wind power, 32, 150, 155, 160, 249, 265, 267, 269, 271–272
Wisconsin, 108, 233, 251
World Bank, 62, 281
world population, 44
World Trade Organization (WTO), 71, 127
World Wildlife Federation, 68
World Wildlife Fund (WWF), 42